1.

Army Dictionary

a

Army Dictionary and Desk Reference

CPT Tim Zurick, USAR

STACKPOLE
BOOKS

Published by
STACKPOLE BOOKS
Cameron and Kelker Streets
P.O. Box 1831
Harrisburg, PA 17105

Printed in the United States of America

Cover design by Caroline Miller

First Edition

10 9 8 7 6 5 4 3 2 1

Library of Congress Cataloging-in-Publication Data

Zurick, Timothy, 1952–
Army dictionary and desk reference / Tim Zurick. — 1st ed.
p. cm.
ISBN 0-8117-2435-2
1. United States. Army—Dictionaries. 2. Military art and science—United States—Dictionaries. I. Title.
UA25.Z86 1992 91-28377
355′.003—dc20 CIP

To my dad, who steered me toward soldiering

Contents

Preface

In most company-level orderly rooms you'll find two basic types of soldiers pushing paperwork. The first is a motivated private or lieutenant with a general understanding of the task but big gaps in his or her knowledge—including the terminology used by veterans or soldiers in other specialties. The resulting misunderstanding can lead to frustration when there's no one around to ask, or to annoyance and the appearance of stupidity when there is. Neither is conducive to mission accomplishment.

The second orderly room denizen you're likely to encounter is a crusty old bird who literally has forgotten the precise difference between "parade rest" and "at ease." He learned it 20 years ago and, frankly, has more important things on his mind—until corrected by a soldier fresh from a Basic course. Finding such tidbits among the ARs and FMs can be daunting.

So we're all faced with the problem of maintaining our skills in communicating in the jargon peculiar to the Army as precisely as possible. This book attempts to catalog the working language of the Army. Although it is not an official publication, it conforms with official definitions. Non Army-specific terms are excluded, as are obscure, archaic, and esoteric terms. I've also attempted to exclude the purely useless: inclusion of all known acronyms, for example, would render this book suitable for use as a big doorstop and not much else. The officially obsolete is occasionally included—simply because it continues in unofficial use.

Thanks are in order to CPT Don Starling and Doreen and Tami Starling for logistical support, and also to Ann Wagoner and Mary Suggs for their kind patience. And especially to Elaine Carlberg for her tireless assistance and long hours at the keyboard.

Branch Acronyms

Each definition is followed by an acronym that represents categorization by branch of usage*, to give a quick orientation for the definition. For example, a definition followed by the acronym MI would indicate that the term is used in the field of military intelligence. The branch acronyms used in this volume are as follows:

ABN Airborne operations, parachuting
AD Air defense, airspace management
ADMIN Administrative functions, information management
ARM Armor, tanks, tracked vehicles, armor plating
ARTY Artillery, long-range guns, targeting
AVN Aviation, air assault, helicopters, air support
CA Civil affairs, military government, public works
COMMS Communications (electronic and non-electronic), message traffic
DC Drill and ceremony
EW Electronic warfare
INF Infantry, small unit tactics, individual combat
INTEL Intelligence
LAW Uniform Code of Military Justice, Law of Land Warfare, Geneva rules
LOG Logistics, materiel support, transportation feeding, supply
MAINT Maintenance
MED Medical services, first aid
MI Military intelligence
MS Military science
NAV Navigation, maps
NBC Nuclear, biological, and chemical warfare
OPFOR Opposing forces, doctrine, weapons
ORD Ordnance, munitions, bombs
PERS Personal equipment, uniforms
SF Special forces and operations

*Note: *The acronym COLL indicates a colloquialism, slang, or jargon.*

STRAT Strategic level, war planning
TAC Tactics, battle planning, and fighting
TNG Training, collective and individual
UW Unconventional warfare, PSYOPs
WPN Weapons (individual and crew-served)

A-10 A fixed-wing, ground support aircraft; the Thunderbolt or Warthog (AVN)

AA **1.** Prosign meaning repeat all after (COMMS); **2.** Antiaircraft (AD); **3.** Assembly area (TAC)

AAA Antiaircraft artillery (AD)

AAFES Army and Air Force Exchange Service (LOG)

AAG Army artillery group (ADMIN)

AALS Active Army Locator System (ADMIN)

AAM **1.** Army Achievement Medal (PERS); **2.** Army aircraft maintenance (AVN)

AATF Air assault task force (AVN)

AB Prosign meaning repeat all before (COMMS)

abatis The use of partially cut, felled trees to obstruct a narrow road (TAC)

ABCA American, British, Canadian, and Australian allies (STRAT)

ABN Airborne (ABN)

abort To discontinue a mission or procedure (COLL)

above the zone Promotion status of officers who have been passed over for promotion once (PERS)

Abrams The main battle tank of the United States; the M1 (ARM)

absolute altitude The altitude of an aircraft above the terrain over which it is flying (AVN)

absolute deviation The distance between the center of a target and a projectile's point of impact (ARTY)

absolute error The distance between the center of a shot group and any single shot in the group (ARTY)

absorbed dose The total amount of nuclear energy retained by an object (NBC)

AC **1.** Active component (ADMIN); **2.** Symbol for the blood agent hydrogen cyanide (NBC)

AC 130 The ground support, gunship version of the C130 Hercules (TAC)

A2C2 Army airspace command and control (AVN)

ACAP Army Child Advocacy Program (ADMIN)

ACC Airspace control center (AVN)

access The authority and ability to obtain knowledge of classified information, based primarily on security clearance level and need to know (INTEL)

accolade Presidential certificate honoring personnel killed or wounded (PERS)

accompanied tour A duty assignment to which dependents are eligible to move (ADMIN)

accountable officer Individual responsible for maintaining property or fund records (ADMIN)

accountable strength The total number of personnel assigned to a unit, regardless of duty status (ADMIN)

accounting classification A fiscal code number indicating the appropriation symbol, allotment number, project account number, object class code, and authorizing fiscal station number (ADMIN)

accounting symbol Combination of letters in a message heading identifying the activity financially responsible for the message (ADMIN)

accredited officer An officer of a foreign government who is authorized access to stipulated classified information (INTEL)

accuracy life The number of rounds for which a weapon is designed to precisely deliver fire (WPN)

accuser Any person who (a) swears to charges, (b) directs that charges be sworn, or (c) has a substantive interest in the prosecution of the charge (LAW)

ACE Armored combat earth mover (ENG)

ACES Army Continuing Education System (TNG)

ACIF Artillery counterfire information (ARTY)

acknowledgment Notification that a message has been received and understood (COMMS)

ACofS Assistant Chief of Staff (ADMIN)

ACR Armored cavalry regiment (ADMIN)

ACS Army Community Services (ADMIN)

action deferred Delay of a tactical operation pending a better tactical situation (TAC)

action officer The officer assigned to ensure that a specific task is accomplished (ADMIN)

activated mine A mine that is fused to detonate if tampered with (WPN)

active Equipment that emits detectable electromagnetic energy (EW)

active air defense Any measures taken to target and destroy enemy aircraft (AD)

Active Army Regular Army troops along with members of the Reserve components on active duty (other than for training) (ADMIN)

active defense The use of limited offensive action/counterattacks to deny the enemy a contested area or objective (TAC)

active duty (AD) Full-time military service (ADMIN)

Active Guard/Reserve (AGR) Reserve component personnel on active duty for more than 180 days in support of Reserve activities (ADMIN)

active homing guidance Target tracking in which the target illuminator and the illumination detector are carried within the missile (WPN)

acute radiation dose Ionizing radiation received over too short a time to permit biological recovery (NBC)

ACV Air cushion vehicle (LOG)

AD Air defense (ADMIN)

ADA Air defense artillery (ADMIN)

ADAM Area denial artillery munition (ORD)

adamsite (ADM) A toxic vomiting agent; decontaminate with individual kit or DS 2 (NBC)

ADC **1.** Assistant division commander (ADMIN); **2.** Aide de camp (PERS)

ADCO Alcohol and Drug Counseling Office (ADMIN)

add Fire adjustment command to increase range (ARTY)

additional duty Unit-level function to be accomplished without a TOE-authorized soldier dedicated against the task (ADMIN)

additional training assemblies (ATA) Inactive duty service authorized for individual reservists to support unit training (TNG)

address group A four-letter code used in routing communications (ADMIN)

ADDS Area data distribution system (ADMIN)

ADF Automatic direction finder (AVN)

ADIZ Air Defense Identification Zone (AD)

adjusted elevation Increase or decrease in a gun's angle (ARTY)

adjusted range Range resulting from adjusted elevation (ARTY)

adjutant (ADJ) An administrative assistant to a commander, responsible for administration and personnel functions (ADMIN)

adjutant general (AG) An administrative assistant to a command with a general staff, responsible for administration and personnel functions (ADMIN)

Adjutant General Corps The combat service support branch and personnel responsible for administrative and personnel functions (ADMIN)

ADM Atomic demolition munition (ORD)

administrative (ADMIN) Nontactical (COLL)

Administrative Adjustment Report (AAR) A document that supports a change in a property balance when the change is otherwise unaccounted for (LOG)

administrative loading Cargo loading of materiel that emphasizes maximum use of space rather than tactical deployment or integrity (LOG)

administrative movement Transportation scheme for troops/vehicles that emphasizes speed and convenience over security; used when enemy contact is considered unlikely (LOG)

administrative net A communications circuit reserved for nonoperational traffic (COMMS)

administrative restriction A commander's limitation of the movement of a member of his command who has been accused of a crime (LAW)

administrative storage The placement of a unit's equipment in a limited care status pending adequate time or supplies to properly maintain it (LOG)

administrative unit A unit capable of managing its own support functions (ADMIN)

ADOA Air defense operations area (AD)

advance by bounds Forward movement of an element in a point-to-point manner, i.e., from cover to cover (TAC)

advance by echelon Forward movement by separate elements of a unit moving successively (TAC)

advanced individual training (AIT) Job-specific training received by enlisted soldiers after basic combat training and before they report to a permanent assignment (TNG)

advance guard An element that precedes a moving force in order to remove obstacles, suppress enemy positions, and exploit any gaps it can detect (TAC)

advance party An element that precedes the main body of a unit for purposes of security, reconnaissance, or arranging for support (ADMIN)

advance section The forward sector of a communications zone (TAC)

advice code Additional instructions from a requisitioner to a supply support activity regarding a supply request (LOG)

ADW Air defense warnings (AD)

ADW Red Code meaning that enemy aircraft and/or missiles are in the area and attack is imminent or in progress (AD)

ADW White Code meaning an attack by enemy aircraft and/or missiles is improbable (AD)

ADW Yellow Code meaning that enemy aircraft and/or missiles are en route and attack is probable (AD)

AEB Aerial exploitation battalion (ADMIN)

AER Army Emergency Relief (ADMIN)

affiliation program A Department of the Army program for preparing Reserve component battalions to integrate with their active component counterparts (ADMIN)

AFQT Armed Forces Qualification Test (PERS)

after action report (AAR) A written review of an operation from an individual's perspective for the purpose of future improvement (ADMIN)

AG Adjutant General's Corps (ADMIN)

agent authentication The provision of personal documents, equipment, and information in support of a covert identity (INTEL)

agility The ability to act more quickly than the enemy to exploit their weaknesses through maneuver of friendly strengths (MS)

AGL Above ground level (AVN)

agonic line A line on a map connecting points of zero magnetic declination (NAV)

AGR Active Guard Reserve (ADMIN)

AHC Assault helicopter company (AVN)

ahkio An individual-drawn sled used for transport of personal/team equipment over snow and ice (INF)

AICV Amphibious infantry combat vehicle (INF)

AID Agency for International Development (STRAT)

aided tracking Alignment of a weapon on a target through use of power equipment, e.g., motors (WPN)

aid man Enlisted soldier attached to a unit for the purpose of emergency medical care or other health service functions (MED)

aid station A treatment facility where limited care and evacuation of sick and wounded are accomplished under the supervision of a medical officer (MED)

aiguillette An ornamental cord worn at the shoulder by an aide or attaché (PERS)

AIM Armored-infantry-mechanized (ADMIN)

aiming circle A device to assist the aiming of a gun by precise measurement of angles or magnetic azimuths (WPN)

aiming point That point on which the sight of a weapon is laid for direction (WPN)

aiming post A marked stake placed in the ground to assist in aligning a weapon on a target (TAC)

air The reported observation of a round's airburst prior to impacting the ground or a target (ARTY)

air adjustment Correction of fire based on air observation (ARTY)

air assault Rapid tactical deployment of infantry troops by helicopter (ADMIN)

airborne (ABN) Parachute deployment of troops, weapons, and/or materiel (TAC)

airborne assault weapon A fully tracked, mobile, air-droppable, antitank gun (WPN)

air-breathing missile A missile that requires an air supply in order to oxidize its propellant fuel, and consequently cannot operate beyond the atmosphere (WPN)

airburst A bomb or projectile that explodes above the ground (NBC)

air cavalry Heliborne combat maneuver force trained to locate the enemy and delay their attack on friendly forces (ADMIN)

air control point A terrain feature or navigational beacon used to regulate air traffic (AVN)

air cushion vehicle (ACV) A vehicle whose motion is supported on a continuously generated column of pressurized air (LOG)

air defense (AD) All countermeasures against enemy aircraft or missiles (TAC)

air defense action area A designated zone of airspace and the ground below it in which friendly aircraft and surface-to-air weapons are given operational priority (TAC)

air defense area Specifically defined airspace for which defense and/or ready control of airborne vehicles must be planned and provided (STRAT)

air defense artillery The combat branch and units responsible for surface-to-air defense (ADMIN)

air defense artillery fire unit The smallest component of an air defense weapon system capable of detecting and destroying a target (AD)

air defense battle zone The specifically designated airspace above and around an air defense fire unit within which aircraft not identified as friendly will be engaged (AD)

air defense identification zone (ADIZ) Specifically defined airspace within which identification, location, and control of airborne vehicles are readily required (AD)

air defense operations area (ADOA) A geographical sector and the airspace above it in which procedures are established to minimize conflict between air defense and other operations; the sector may be further subdivided as follows: (a) air defense action area, (b) air defense area, (c) air defense identification zone, and (d) firepower umbrella (STRAT)

air defense sector A tactical subdivision of an air defense area (AD)

air defense warning conditions A system for specifying the probability of enemy air attack; *See* ADW White, ADW Yellow, ADW Red (AD)

air defense weapons control status Degrees of restriction over the use of air defense weapons in a combat zone; *See* weapons free, weapons hold, weapons tight (AD)

airdrop Delivery of troops or materiel from an aircraft in flight (LOG)

air ground operations system The system by which the ground commander receives, coordinates, and forwards requests for air support and disseminates related intelligence (ADMIN)

airhead The area around a landing zone that must be secured to maintain continuous air landing operations (TAC)

airland To land with the aircraft rather than by parachute (ABN)

AirLand battle Current Army doctrine describing (*See* FM 100-5, "Operations") how modern warfare will be prosecuted: on a nonlinear battlefield, extended in depth by time, resources, and terrain, with integration of nuclear and chemical weapons and close coordination between air and ground forces (STRAT)

AirLand operation The air movement of personnel and materiel to an airfield for further deployment (LOG)

air liaison officer An aviator attached to a ground unit to advise that commander on air support/operations (ADMIN)

air lock Small, sealed area between a contaminated environment and a clean shelter area, from which contamination can be flushed after personnel pass through (NBC)

Air Medal Decoration for meritorious achievement in flight (PERS)

air mobile operation The use of heliborne combat power in support of ground operations (TAC)

air photographer Soldier trained to interpret air reconnaissance photography and assemble the photographs into mosaics and maps (INTEL)

air signal Air-to-ground communication by aircraft maneuver when radio communication is not possible (COMM)

air spot Adjustment of gunfire by air observation (ARTY)

air support The use of airborne weapons to assist a ground battle (TAC)

air trooping Administrative air movement of soldiers (LOG)

AIT Advanced individual training (TNG)

ALAT Army Language Aptitude Test (TNG)

ALCANUS Alaska, Canada, United States (STRAT)

alert A warning to troops to stand by for action (TAC)

alibi Leftover round (due to temporary weapon malfunction) fired on a range after the main volley of other firers is expended (TNG)

ALICE All-purpose, lightweight individual carrying equipment (INF)

all available A request or order to utilize total artillery assets against a target (ARTY)

all clear Signal that the danger of immediate attack has passed (COMMS)

allotment Portion of a soldier's pay deducted and paid out to a third party (ADMIN)

ALO **1.** Air liaison officer (PERS); **2.** Authorized level of organization (ADMIN)

alternate position Tactical position to be occupied when the primary position becomes unusable; location must be capable of supporting the original mission (TAC)

alternate traversing fire Coverage of an area target in width by changing the gun's deflection and taking advantage of normal range variation between rounds to cover the target in depth (ARTY)

ambush A surprise attack from a concealed position upon an enemy in transit (TAC)

AMCITS American citizens (CA)

AMDF Army Master Data File (LOG)

AMG Antenna mast group (COMMS)

AMLS Airspace management liaison section (AVN)

amnesty box Container for the deposit of contraband with no questions asked, prior to a shakedown inspection (LAW)

amphibious operation A surface attack launched from naval craft against a hostile shore (TAC)

amphibious tank Tank capable of delivering direct or indirect fire from shallow water as well as ashore (ARM)

amplification Message text that expands on only one preceding information set (COMMS)

amplitude modulation (AM) Variation in the positive/negative voltage value of a broadcast radio signal in order to enable it to carry information (COMMS)

AMSA Area Maintenance Support Activity (LOG)

analysis The stage in the intelligence production cycle in which raw information is refined into intelligence by examining significant facts and drawing conclusions from them (INTEL)

anchor cable A metal line within an aircraft to which parachute static lines are attached (ABN)

ANCOC Advanced Noncommissioned Officers' Course (TNG)

ANG Air National Guard (ADMIN)

angle of site The vertical angle between the base of the trajectory and the straight line joining the origin and the target (ARTY)

angle of traverse Horizontal range of motion of a gun on its mount (WPN)

angle T The angle formed by the intersection of the gun-target line and the observer-target line (ARTY)

angular travel The distance away from line of sight traveled by a moving target in a specified time (ARTY)

annex **1.** Reference material attached to a main document (ADMIN); **2.** A support building (LOG)

annual training (AT) The minimum period (usually two weeks) that a Reserve component soldier must perform active duty each year (TNG)

antenna An electrical conductor that radiates or receives electromagnetic waves (COMMS)

antenna pattern A diagram showing the intensity of signal strength in different directions from a broadcast antenna (COMMS)

antidim A substance that reduces fogging of the eyepieces of a protective mask (NBC)

antilift device A trigger designed to detonate a mine if it is tampered with (WPN)

antimaterial agent A chemical or living organism that destroys or damages selected material (WPN)

antiradiation missile A missile that passively homes on a radiation source (WPN)

ANZUS Australia, New Zealand, and the United States (STRAT)

AO Area of operations (TAC)

AOAP Army Oil Analysis Program (LOG)

AOE Army of Excellence (ADMIN)

Apache The AH-64, a rotary-wing, armed attack aircraft (AVN)

APC Armored personnel carrier (INF)

APFT **1.** Army physical fitness training (TNG); **2.** Army physical fitness test (TNG)

APICM Antipersonnel improved conventional munitions (ORD)

APO Army Post Office (ADMIN)

apex angle Horizontal angle between a line from the target to a gun and a line from the target to the observation post (ARTY)

appreciations Those facts, estimates, and assumptions regarding an enemy that form a basis for planning and decision making (MS)

approach march Manner of advance to an attack position when enemy contact is expected (TAC)

approved forces Forces authorized in the Five-Year Defense Program of the Secretary of Defense (STRAT)

appurtenance A device representing an additional award of a decoration (PERS)

APRT Army Physical Readiness Test (PERS)

APU Auxiliary power unit (LOG)

AR **1.** Army Regulation (ADMIN); **2.** Prosign meaning end of transmission (COMMS)

ARC Accounting requirement code (LOG)

ARCAM Army Reserve Components Achievement Medal (PERS)

ARCOM **1.** Army Commendation Medal (PERS); **2.** United States Army Reserve Command (ADMIN)

area command Military authority over all services within a geographic region (ADMIN)

area of influence Geographical sector that a commander is capable of directly affecting through fire and maneuver of his or her unit (TAC)

area of interest A commander's area of influence, plus adjacent areas and objective areas within enemy territory (TAC)

area of operations (AO) A sector in which combatants are deployed (STRAT)

area of probability An area containing a fix, the exact position of which is undetermined (EW)

area of responsibility Sector entrusted to a commander for the purpose of controlling movement and tactical operations and developing and maintaining installations (TAC)

area search Reconnaissance of a specified sector to obtain information on activities there (INTEL)

area study A report on all factors that form the background for an unconventional warfare scenario (UW)

ARFCOS Armed Forces Courier Service (ADMIN)

arm **1.** Any branch of the Army with a combat or combat support mission (ADMIN); **2.** Antiradiation missile (WPN); **3.** To enable for use or detonation by removing a safety mechanism or aligning explosive elements in an explosive train (WPN)

armament error The dispersion of shots from a single gun not due to personnel error or weapon adjustment (ARTY)

Armed Forces Courier Service (ARFCOS) A joint service providing secure and expeditious transmittal of sensitive or classified military documents and/or materiel (COMMS)

armed forces intelligence The component of strategic intelligence dealing with the military capabilities of foreign nations (INTEL)

Armed Forces Qualification Test (AFQT) Diagnostic measure of a soldier's mental ability and potential (ADMIN)

Armed Forces Reserve Medal Award in recognition of 10 years of honorable service within a 12-year period in any Reserve component (PERS)

armed reconnaissance A mission with the dual purpose of finding and attacking targets of opportunity (TAC)

arming The conversion of a weapon or munition from a safe condition to readiness for use (WPN)

arming range The distance from a gun at which a fuse enables a round to detonate (ARTY)

armor The combat branch and units consisting of tanks and armored cavalry/reconnaissance (ADMIN)

armored cavalry Combat units characterized by high mobility, shock action, and flexible communications; designed to execute reconnaissance, security, or economy of force operations (TAC)

armored infantry A unit designed to engage and destroy the enemy by fire and maneuver, to repel assault in close combat, and to provide support for tanks (STRAT)

armored personnel carrier (APC) A lightly armored, tracked, highly mobile troop transport for use in tactical operations (TAC)

armored reconnaissance airborne assault vehicle The Sheridan, a lightly armored, air-droppable tracked vehicle that functions as the principal assault weapon of airborne units (ABN)

armorer Soldier who maintains a unit's small arms (LOG)

army The land-based military forces of a nation (STRAT)

Army airground system The staff and communications used to manage coordination of air support and ground operations (ADMIN)

Army Achievement Medal Decoration made in recognition of distinguished or meritorious noncombat service of a lesser degree than that required for the Army Commendation Medal (PERS)

Army classification battery Aptitude tests designed to measure soldiers' potential proficiency in the various military occupational specialties (TNG)

Army Commendation Medal Decoration made in recognition of heroism and meritorious achievement in a lesser degree than those recognized by the Bronze Star or Soldier's Medal (PERS)

Army Commitment Board A committee formed to determine the sanity of a soldier and whether the soldier should be institutionalized (ADMIN)

Army General Staff Those specialized officers who render advice and assistance to the Secretary of the Army and his assistants (ADMIN)

Army group Two or more field armies combined under a designated commander (ADMIN)

Army master data file (AMDF) The basic information used to manage and distribute supplies and materiel (LOG)

Army Medical Department Those branches that report to the Surgeon General, including the Medical Corps, the Dental Corps, the Veterinary Corps, the Medical Service Corps, the Army Medical Specialist Corps, and the Army Nurse Corps (ADMIN)

Army National Guard (ANG) The army component of the militia of the United States and its various districts (ADMIN)

Army Reserve Command (ARCOM) A headquarters designed to command a grouping of nondivisional units within a given area (ADMIN)

Army service area The sector between the corps rear and the combat zone rear; an administrative area (ADMIN)

Army Stock Fund The system for managing inventories of materiel belonging to the Army (LOG)

Army Subject Schedule A centralized DA publication providing detailed guidance on presentation of training in branch, general, or MOS subjects (TNG)

Army Training and Evaluation Program (ARTEP) A series of publications that provide collective training guidance based on unit type (TNG)

ARNG Army National Guard (ADMIN)

arrowhead Device worn on the Asiatic-Pacific Campaign, European-African-Middle Eastern Campaign, Korean Service, and Vietnam Service ribbons with the point facing up and to the wearer's right of service stars; one per ribbon (PERS)

ARTEP Army Training and Evaluation Program (TNG)

Article 15 Section of the Uniform Code of Military Justice that provides for swift, nonjudicial punishment of minor offenses (ADMIN)

artificial daylight Illumination of a nighttime battlefield to a brightness lighter than that created by a full moon (TAC)

artificial moonlight Illumination of a nighttime battlefield to a brightness lighter than starlight but not as light as the full moon (TAC)

artillery The combat branch and units based on large-caliber (30 mm) guns, rockets, and missiles and their employment (ADMIN)

ARTY Artillery (ADMIN)

ARTYMET Artillery meteorological team (INTEL)

AS Prosign meaning wait (COMMS)

ASAP As soon as possible (COLL)

ASI Additional Skill Identifier (PERS)

ASL Authorized stockage level (LOG)

ASM Air-to-surface missile (AD)

ASP **1.** Ammunition supply point (LOG); **2.** All-source production (INTEL)

ASPS All-source production section (INTEL)

assault The final stage of an attack; characterized by direct or hand-to-hand combat (TAC)

assault fire Short-range fire delivered by attacking units on point targets (TAC)

assault footpath A one-meter-wide path through a minefield; cleared to permit dismounted troops to secure the minefield's far side (TAC)

assault force The element responsible for seizure of the objective area in an offensive operation (TAC)

assault gun Self-propelled or tank-mounted weapons used for direct fire against point targets (WPN)

assault phase of military government The period that begins with first contact between ground forces and local civilians and continues through establishment of control by the ground forces (CA)

assault position A position between the line of departure and the objective to which the assault echelon deploys prior to engaging the enemy (TAC)

assault wire Light, man-portable communications wire suitable for tactical conditions (COMMS)

assembly area A site at which a command is formed in preparation for an operation (TAC)

asset **1.** An item or unit of military value (MS); **2.** A source of military information (INTEL)

assign To place under the permanent control of another command (ADMIN)

astigmatizer A device that, through a range finder, enhances observability of objects at night (WPN)

astronomical twilight That time of incomplete darkness when the center of the sun is 18 degrees below the horizon (ADMIN)

ASVAB Armed Services Vocational Aptitude Battery (ADMIN)

asylum Safe haven from conflict granted by a nation to refugees, prisoners of war, or troops who have deserted the conflict (LAW)

as you were Command that cancels a preparatory command; also used in lieu of saying "I made a mistake" (COLL)

AT **1.** Annual training (TNG); **2.** Antitank (WPN)

ATA **1.** Alternate training assembly (TNG); **2.** Actual time of arrival (ADMIN)

ATACMS Army Tactical Missile System (ARTY)

ATACS Army Tactical Area Communications System (COMMS)

A team The basic Special Forces operational detachment of 12 or 13 specialized soldiers (UW)

at ease **1.** Combined command that directs a formation of soldiers to keep at least their right foot in place and remain silent; otherwise, they may move or relax (DC); **2.** Be quiet (COLL)

ATGM Antitank guided missile (WPN)

ATI Artillery target intelligence (ARTY)

ATKHB Attack helicopter battalion (ADMIN)

ATP Allied Technical Publications (ADMIN)

atropine A chemical used to counteract the effects of a nerve agent (MED)

ATS Automated telecommunications system (COMMS)

attach To temporarily place personnel or units under military command and control; administrative functions (i.e., transfer, promotion) are normally retained by the home unit (ADMIN)

attaché A soldier whose duties include diplomatic assistance functions on military matters (PERS)

attack A movement to engage and destroy the enemy (MS)

attack position The last position short of the line of departure occupied by assault elements (TAC)

attention Combined command (spoken in two parts) directing soldiers to assume a silent stance with head and eyes forward, heels together with feet at a 45-degree angle, body erect, legs straight, and arms hanging without stiffness at the sides (DC)

attention to orders An adjutant's command announcing that he or she is about to read orders (DC)

attenuation Decrease in signal strength due to absorption and scattering (COMMS)

attrition Reduction in combat effectiveness due to loss of personnel and materiel (MS)

AUEL Automated unit equipment list (LOG)

augmentation Reinforcement of a command through redeployment of forces from other commands (ADMIN)

aural null The lack of an audible signal that occurs when a directional antenna is aligned perpendicular to the direction of a signal source (EW)

AUS Army of the United States (ADMIN)

authentication Determination that a communication is from a legitimate friendly source and not the result of enemy intrusion (COMMS)

authenticator The enciphered portion of a message that validates its legitimacy (COMMS)

authorized Officially sanctioned or required (ADMIN)

authorized level of organization (ALO) The percentage of assets (personnel and equipment) that a unit is authorized to maintain compared to its complete table of organization (ADMIN)

authorized stockage list (ASL) Those items to be kept on hand at a specific supply echelon (LOG)

authorized stoppage Any pay withheld from a soldier's check, whether voluntary or involuntary (PERS)

authorized strength The maximum number of personnel spaces authorized for a command, without regard to whether they are to be filled at a given time (ADMIN)

AUTODIN Automatic digital network (COMMS)

autofrettage Stress hardening of a cannon tube (ARTY)

automatic Gun that continues to chamber and fire rounds as long as the trigger is held (WPN)

automatic return The mandatory return of an item to its national inventory control point for required repair (LOG)

AUTOSEVOCOM Automatic, secure, voice communications (COMMS)

AUTOVON Automatic voice network—the basic military telephone system (COMMS)

auxiliary Organized civilian support for a resistance force (UW)

auxiliary target A position on which fire can be adjusted and then shifted to an actual target, the intent being to surprise the actual target (ARTY)

aviation The combat branch and units responsible for employment of the Army's air assets (ADMIN)

AVLB Armored, vehicle-launched bridge (LOG)

AVN Aviation (ADMIN)

AVUM Aviation unit maintenance (AVN)

AWACS Airborne warning and control system—a long-range radar air platform (INTEL)

AWADS Adverse weather aerial delivery system (ABN)

award Citation or recognition for exceptional acts or services (ADMIN)

AWOL Absent without leave (ADMIN)

axis of advance The main path of an offensive operation, usually a cleared area or road network (TAC)

axis of trunnions Axis around which a gun's elevation is rotated (ARTY)

azimuth A direction, expressed as an angle, measured clockwise from a reference point (NAV)

B

back azimuth The direction opposite an azimuth—an azimuth plus or minus 180 degrees (NAV)

back blast The rearward explosive force of gases from certain recoilless weapons and rocket launchers (WPN)

back channel An unofficial source of friendly information or means of communication (COLL)

back tell Communication of information from a higher to a lower command echelon (INTEL)

BAI **1.** Basic area of interest (INTEL); **2.** Battlefield air interdiction (TAC)

balaclava A cold weather ski mask (PERS)

balisage Nighttime marking of a route with dim lighting to maintain blackout conditions (LOG)

ball ammunition A small arms cartridge with a solid core bullet for use against targets not requiring armor piercing capability (WPN)

ballistic missile A self-propelled projectile without aerodynamic lifting surfaces that consequently follows a ballistic trajectory upon termination of thrust (WPN)

ballistics Those factors that affect the flight of a projectile: muzzle velocity, weight, size, shape, rotation of the earth, and atmospheric conditions (ARTY)

ballistic temperature A computed temperature with the same effect on a projectile as the actual temperature it will encounter (ARTY)

ballistic trajectory The path of a projectile after the propulsive force is terminated (WPN)

band **1.** Two or more lines of mutually supporting physical obstacles (TAC); **2.** A range of radio wavelengths (COMMS)

bandolier A cloth loop with pockets for carrying small arms ammunition; worn over the shoulder (INF)

bangalore torpedo A field-expedient, high explosive charge used in clearing operations (e.g., of minefields or obstacles) (ORD)

barrage A barrier of gunfire (TAC)

barrage jamming Transmission of electromagnetic energy meant to disrupt a broad range of radio frequencies (EW)

barrel Tube that gives initial direction to a fired projectile (WPN)

barrel erosion The wearing down of a bore's surface due to the mechanical abrasion of the rounds, explosive gas pressure, and cleaning (WPN)

BAS **1.** Basic allowance for subsistence (ADMIN); **2.** Battlefield automated systems (ADMIN)

BASD Basic active service date (ADMIN)

base defense Measures to prevent sabotage and direct attack on a military installation (TAC)

base-level commercial equipment Nonstandard, non-centrally managed materiel that is authorized by tables of distribution and allowances and costs $3,000 or more (LOG)

base map A graphic representation of terrain that contains minimal other information and is designed with the intent of being overlaid with various types of additional information (NAV)

base of fire Fire that supports attack operations of maneuver elements (ARTY)

base piece The gun in a battery that serves as the basis for calculating initial firing data (ARTY)

base unit Organization around which a tactical operation is planned (ADMIN)

basic allowance for quarters (BAQ) Money paid to all service members who are not provided with housing (ADMIN)

basic allowance for subsistence (BAS) Monetary payment issued to soldiers authorized to purchase their own food (ADMIN)

basic branch The area of military specialty to which a new officer is assigned upon commissioning (ADMIN)

basic combat training Fundamental training in military topics, tactics, and discipline given to newly enlisted soldiers (TNG)

basic end item A major assembly or system that can be divided into component subunits and issued to lower echelons (LOG)

basic issue items Those items of equipment distributed along with the major end items they support; e.g., a tool kit issued with a tank (LOG)

basic load That quantity of expendable supplies authorized to be on hand, per unit or soldier, to sustain combat until routine resupply can be delivered (LOG)

basic unit training Initial training of a soldier in the role he or she will assume upon his or her permanent duty assignment (TNG)

basis of issue Document that authorizes the distribution of supplies and equipment (LOG)

BAT-D Deception battalion (ADMIN)

battalion A unit at a command level below a brigade and above a company/battery (ADMIN)

battery An artillery unit at a command level below a battalion and above a platoon; equivalent to a company or troop (ADMIN)

battery front Distance between a battery's flank guns (ARTY)

battery left (or right) Command to discharge weapons in sequence, from the specified flank, at five-second intervals; also troop left (or right) (ARTY)

battle focus Emphasis on a unit's wartime mission in planning of its training (TNG)

battle map A map of sufficient detail to show tactical terrain use for all forces; normal scale: 1:25,000 (TAC)

battle sight Calibration of a weapon sight to cause a round to hit a target at its center of mass at an anticipated combat range (WPN)

bay Section of a floating bridge (LOG)

BCC Battery control central (ARTY)

BCS Battery computer system (ARTY)

BCT Basic combat training (TNG)

BDE Brigade (ADMIN)

BDO Battle dress overgarment (NBC)

BDU Battle dress uniform (PERS)

beacon An emitter of electronic energy used as a navigational aid (NAV)

beam rider A missile that tracks on an electronic beam (AD)

bearing Direction from one point to another; measured in degrees of horizontal, clockwise angle (NAV)

beaten zone The area on the ground where the rounds of a cone of fire fall (WPN)

beehive An antipersonnel projectile, loaded with fléchettes (ORD)

belay To secure by tautening the loose end of a rope down which another soldier is rappelling (TAC)

below the zone Promotion status of officers who are in the year group behind those under primary consideration for advancement to the next grade (COLL)

bench mark A position of known location and elevation from which calculations can be made regarding other positions (NAV)

BEQ Bachelor enlisted quarters (LOG)

berm An artificial ledge or shoulder of ground; built to deflect fire (TAC)

bessel method The location of one's position on a map by comparison to visible terrain features (NAV)

beta radiation Nuclear emissions with a range of 10 to 15 meters; capable of producing skin burns (NBC)

BG Brigadier general (PERS)

BI **1.** Background investigation (INTEL); **2.** Battlefield illumination (TAC); **3.** Branch immaterial (ADMIN)

BII Basic issue items (LOG)

billet **1.** Quarters, housing (LOG); **2.** A personnel authorization slot (ADMIN)

binary weapon A weapon (explosive or chemical) that is stored in two harmless components and armed only by assembly (WPN)

biological agent A microorganism used in weapons to cause disease in personnel or deterioration of materiel (NBC)

bird **1.** A full colonel (COLL); **2.** An aircraft (COLL)

BITE Built-in test equipment (MAINT)

biting angle Smallest angle at which a projectile will penetrate armor rather than deflect away (ORD)

bivouac A temporary encampment of tents or improvised shelters (TAC)

BL Basic load (LOG)

black Designation for an illegal intelligence mission, rather than one operating under an apparently legitimate cover (INTEL)

black book A chronological compilation of classified messages (INTEL)

black designation Circuitry carrying message traffic that is either encrypted or unclassified (COMMS)

black forces NATO designation for opposing forces within OPFOR exercises (INTEL)

black hat A member of the training cadre at the airborne school, Fort Benning, Georgia (ABN)

Blackhawk The UH-60A rotary-wing, combat assault transport and electronic warfare and target acquisition aircraft (AVN)

blacklist A listing of confirmed or suspected civilian enemy collaborators whose potential actions threaten the security of friendly forces (INTEL)

black powder An unstable, sensitive, low explosive, general purpose charge (WPN)

black propaganda Propaganda falsely sourced—created by one nation and attributed to another (UW)

blank ammunition Rounds that contain no projectile but only a charge to produce sound for simulation/training purposes (ORD)

blasting cap Small container filled with a sensitive primer used to detonate a large, more stable explosive charge (ORD)

blasting machine A small generator used to energize an electrical firing circuit (WPN)

blast wave A wave of sharply increased air pressure resulting and radiating from a detonation (WPN)

blind circuit Channel in which communication is possible in one direction only (COMMS)

blind transmission Message sent without expectation of acknowledgment (COMMS)

blister agent A chemical weapon that creates lesions on any body surface it contacts (NBC)

blivet A large, collapsible storage bladder for liquids, especially fuel (LOG)

blood agent A chemical that prevents the normal transfer of oxygen between blood and body tissue (NBC)

blood chit A card offering a reward to anyone assisting the bearer of the card to safety (CA)

blowback The force of rapidly expanding gases resulting from the firing of a weapon (WPN)

blue bark A report concerning movement and treatment of family members of deceased servicemembers (ADMIN)

blue bell A report concerning suspected criminal conduct or mismanagement that could damage public confidence in the Army (ADMIN)

blue forces The designation for friendly forces during NATO exercises (TNG)

BMCT Beginning morning civil twilight (ADMIN)

BMD 1. Soviet-bloc airborne amphibious infantry combat vehicle (OPFOR); 2. Ballistic missile defense (AD)

BMG Budget and manpower guidance (ADMIN)

BMNT Beginning morning nautical twilight (ADMIN)

BMP Designation for a series of Soviet-bloc armored amphibious infantry combat vehicles (OPFOR)

BN Battalion (ADMIN)

BNCOC Basic Noncommissioned Officers' Course (TNG)

bolo A training failure (TNG)

bona fide Anything that tends to establish an agent's identity/loyalty (INTEL)

booby trap An explosive device designed to produce casualties upon the disturbance of an object that appears harmless (WPN)

BOQ Bachelor officers' quarters (LOG)

border crosser A civilian living near an international boundary who routinely crosses over for legitimate purposes (LAW)

bore The inside diameter of a gun's barrel (WPN)

boresafe fuze A device that arms a round only after it has cleared the muzzle of a weapon (ORD)

boresighting The aligning of the bore of a weapon and the aiming device used to fire it (WPN)

bouncing Betty A bounding mine (WPN)

bounding mine Antipersonnel mine that, for maximum effect on troops, blows itself up to a height of three to four feet prior to detonation of its main charge (WPN)

bounding overwatch Form of movement in which one element moves quickly while the other provides security from a covered/concealed position—they alternate in coverable bounds; used when probable enemy contact takes precedence over speed (TAC)

bourrelet The surface of a round on which it slides through a gun tube (ORD)

BPED Basic pay entry date (ADMIN)

BPS Basic psychological operations study (UW)

bracketing To adjust fire by firing to either side of a target and successively halving the direction of aim between the original points of impact until the target is hit (ARTY)

Bradley The M2 or M3 infantry fighting vehicle/armored personnel carrier (INF)

branch Area of specialization for officers (ADMIN)

branch immaterial position A duty position that can be filled by a commissioned officer of any specialty (ADMIN)

branch material position A duty position that can be filled only by a commissioned officer of a specific specialty (ADMIN)

brassard Distinctive temporary insignia worn on the shoulder to identify the wearer to the public or to designate the wearer for a special task (ADMIN)

breaching Expedient passage through a fortification or an obstacle (TAC)

breakout Maneuver in which encircled forces defend in all directions but one, in which they launch an offense to breach the encirclement and lead all forces to safety (TAC)

breastwork An above-ground, defensive earthwork that provides cover for standing riflemen (TAC)

breech The back opening of a gun (WPN)

breechblock Steel block that closes the back end of a cannon (WPN)

brevity code Use of abbreviated words or phrases to reduce radio transmission time (COMMS)

bridgehead The area to be secured around a bridge in order to maintain its continuous use (TAC)

briefback An oral presentation to a command element of a mission plan in response to the command's stated objectives (ADMIN)

briefing An orally presented summary of a current situation (ADMIN)

brigade (BDE) A unit at a command level below a division and above a battalion (ADMIN)

brigadier general (BG) A commissioned officer; pay grade O–8 (PERS)

broadcast method A communication network in which one station transmits and the rest merely receive without acknowledgment (COMMS)

Bronze Star Medal Decoration in recognition of heroic or meritorious service in connection with armed conflict operations to a lesser degree than that meriting the Silver Star or the Legion of Merit (PERS)

BSA Brigade support area (ADMIN)

BT 1. Prosign meaning break text (COMMS); 2. Basic training (ADMIN)

BTB Blind transmission broadcast (COMMS)

BTMS Battalion training management system (TNG)

BTR Designation for a series of Soviet-bloc armored personnel carriers (OPFOR)

BUCS Backup computer system (ADMIN)

buddy system Deployment or training of soldiers in pairs for the purpose of mutual support (ADMIN)

bunker A fortified defensive position for weapons and/or personnel (TAC)

burn To expose the identity of an undercover agent (INTEL)

burn bag A container for classified waste (INTEL)

burster An explosive charge that breaks open chemical projectiles and bombs (ORD)

burster course Layer in a fortification designed to detonate projectiles before their explosive force can fully penetrate the fortification (TAC)

bursting charge An explosive charge designed to break open a projectile (ORD)

burst range The horizontal distance from an artillery piece to the point of burst of one of its rounds (ARTY)

burst transmission Broadcast of an electronic message in condensed form to minimize time of transmission (COMMS)

bust An abnormality in an encrypted message text that is of value in attempting to decrypt (INTEL)

butterbar A second lieutenant (COLL)

butt stroke To strike with the stock section of a rifle (INF)

bypass To maneuver around one flank of an enemy, avoiding engagement (TAC)

by the numbers 1. Preparatory command indicating that the command of execution will be broken down into its steps and controlled by the speaker (DC); 2. To do something exactly as prescribed by regulation (COLL)

C

C 1. Prosign meaning correct (COMMS); 2. Change to a gun's elevation that will result in a range correction of 100 meters (ARTY)

C² Command and control (ADMIN)

C³ Command, control, and communications (ADMIN)

C-5A The fixed-wing Galaxy, largest transport aircraft in the U.S. inventory (AVN)

C-130 The Hercules, a fixed-wing, medium-range transport aircraft (AVN)

C-141 The Starlifter, a fixed-wing, long-range transport aircraft (AVN)

CA Civil affairs (ADMIN)

CAB Combat aviation brigade (ADMIN)

cable block Road obstruction consisting of a steel line stretched diagonally across a road (TAC)

cadastrial map A map of such large scale that the exact position of certain objects is shown (NAV)

cadence The rhythmic pace at which movement in formation is executed (DC)

cadet (CDT) An officer-in-training (TNG)

cadre Nucleus of key, permanent party personnel (ADMIN)

caduceus The symbol of the Medical Corps: serpents entwined around a winged staff (MED)

CAE Criterion action element (TNG)

calcium hypochlorite Chemical used to purify water (LOG)

CALFEX Combined arms live fire exercise (TNG)

caliber (CAL) The operating diameter of a bullet or shell and the bore of the gun used to fire it (WPN)

call fire Fire delivered in response to a specific request for support (ARTY)

call sign An encrypted identification for a given radio transmitter station (COMMS)

call up The identifying signals by which stations contact each other (COMMS)

camouflage The visual disguise of potential targets to make them blend in with their surroundings (TAC)

camp A semipermanent base for troops—less permanent than a fort, but more permanent than a bivouac (LOG)

canalize To force enemy elements into a narrow, elongated killing zone by employment of obstacles and fire (TAC)

C & J Collection and jamming (INTEL)

cannelure A groove that attaches a cartridge case to a bullet (ORD)

cannibalization The use of parts from deadlined equipment to temporarily maintain the operation of similar systems (LOG)

canopy The portion of a parachute that fills with air to slow descent (ABN)

cant The tilt of the gun trunnion axis resulting from a tank being on uneven ground (ARM)

cantonment Temporary shelter for troops or materiel (LOG)

cantonment area The part of a post that includes offices, barracks, housing, and other support facilities (ADMIN)

CAO Casualty assistance officer (ADMIN)

Capabilities exercise (CAPEX) An exercise designed to show the capabilities of several weapon systems through a live fire demonstration (TNG)

CAPEX Capabilities exercise (TNG)

captain A commissioned officer; pay grade O-3 (PERS)

capture tag Label affixed to enemy prisoners of war describing the time, date, and place of capture and other important circumstances regarding the capture (INTEL)

Career management field (CMF) The grouping of related MOSs to enhance the management of soldiers' careers while better meeting assignment needs of units and installations (ADMIN)

cargo carrier A tracked, highly mobile materiel transporter designed to resupply maneuver units (LOG)

CARP Computed air release point (ABN)

carriage A support for a large gun that may also include elevating and traversing mechanisms (WPN)

carry light Searchlight used to maintain illumination of a target until it can be fired upon (TAC)

CARS Combat arms regimented system (ADMIN)

cartridge An assembly of the propellant, projectile, and any casing that comprises one round or shot from a weapon (ORD)

CAS Close air support (TAC)

CAS^3 Combined arms and services staff school (TNG)

casual Status of personnel awaiting orders or duty assignment; currently unassigned (COLL)

casual payment Salary or expense money paid to a soldier in advance of an exact determination of his or her full entitlement (ADMIN)

casualty A soldier involuntarily lost from duty status (ADMIN)

Casualty assistance officer (CAO) The senior NCO or officer charged with helping the surviving family members of a soldier killed in the line of duty (ADMIN)

cat eyes Any luminescent or dimly illuminated devices affixed to an object or person to enhance visibility under blackout conditions (COLL)

cat hole A shallow hole dug as a field-expedient latrine (TAC)

caveat Army general restriction on access to classified information (e.g., NOFORN—not releasable to foreign nationals) (INTEL)

CAVU Ceiling and visibility unlimited (AVN)

CBS-X Continuing balance system—expanded (LOG)

CCA Contamination control area (NBC)

CCF **1.** Correctional custody facility (ADMIN); **2.** Central clearance facility (INTEL)

C-day The date on which deployment for an operation commences; a planning point within the wartime manpower planning system (ADMIN)

CDM Chemical downwind message (NBC)

CE Communications electronics (COMMS)

cease engagement A fire control order directing a battery to stop tracking a particular target; missiles already in flight may be permitted to continue (AD)

cease fire A fire control order directing a battery to refrain from firing on but to continue tracking targets; missiles already in flight may be permitted to continue (AD)

ceiling The lowest altitude at which visibility is obscured (AVN)

CENTAG Central Army Group, Central Europe (ADMIN)

center of impact The middle point, in range and direction, within the dispersion pattern of impact bursts (ARTY)

centigray The unit of absorbed radioactivity; equal to 100 ergs of energy per gram (NBC)

CENTO Central Treaty Organization (STRAT)

Central Intelligence Agency (CIA) Organization with ultimate responsibility for all forms of national U.S. intelligence (INTEL)

central purchase Acquisition of an item of materiel through a single authority for distribution throughout the supply system (LOG)

CEOI Communications–electronics operation instructions (COMMS)

certificate of capacity Statement of an officer's qualification for promotion (ADMIN)

certificate of expenditures Statement that an expendable item has been consumed through use and should therefore be dropped from accountability (LOG)

certificate of honorable service Recognition of a deceased soldier's service (ADMIN)

CEV Combat engineer vehicle (ENG)

CEWI Combat electronics warfare intelligence (EW)

CF Copy or copies furnished (ADMIN)

CFA **1.** Current files area (ADMIN); **2.** Covering force area (TAC)

CFL Coordinated fire line (TAC)

CFX Command field exercise (TNG)

CFZ Critical friendly zone (ARTY)

CG **1.** Symbol for the choking agent phosgene (NBC); **2.** Commanding general (ADMIN)

CGS Command and general staff (ADMIN)

cGy Centigray (NBC)

chaff Strips of metallic material distributed in the atmosphere to disrupt radar signals (TAC)

chain of command The organizational structure through which military authority is exercised (ADMIN)

challenge A secret word, to which an unknown person must reply with a password that only an authorized person should know (TAC)

CHAMPUS Civilian Health and Medical Program of the Uniformed Services (MED)

Chapparal The M48 tracked, mobile, short-range, low-altitude, air defense missile system (AD)

Chaplain Corps The combat service support branch and personnel responsible for providing religious ministration to soldiers (ADMIN)

chaplain fund A nonappropriated money account used in support of activities related to religious programs of a command (ADMIN)

Chaplains' Activities Fund An appropriated money account for the purchase of religious materiel (ADMIN)

charge A quantity of explosive used for propulsion of a projectile or demolition (ORD)

chart direction of wind Horizontal, angular difference between the azimuth of fire and the ballistic wind azimuth (ARTY)

chatter Unnecessary radio conversation (COMMS)

check fire A command to temporarily halt firing (ARTY)

checkpoint A site at which vehicular or pedestrian traffic is stopped and inspected (ADMIN)

checksum digit The last numeral of the sum of a sequence of numerals; for example, the checksum digit of 3911 is 4 (3 + 9 + 1 + 1 = 1*4*) (COMMS)

chemical agent A substance intended to kill, injure, or incapacitate a soldier through its physiological impact (WPN)

Chemical Corps The combat support branch and soldiers responsible for providing support and expertise in matters of nuclear, biological, and chemical warfare (ADMIN)

chemical downwind message (CDM) A warning issued every six hours as to the degree of hazard to be expected resulting from the scattering of chemical agents by predicted wind (NBC)

chemical survey Evaluation of an area to determine the nature, distribution, and severity of the potential hazard from chemical agents (NBC)

chemwarn Message warning of a friendly chemical strike (NBC)

chevron Upside-down V-shaped insignia of enlisted rank (PERS)

chief of staff The principal assistant to a commander at brigade level or higher (ADMIN)

chilblain Damage to skin caused by prolonged exposure to cold air; symptoms include swelling, itching, infection, and bleeding; treat by gradual rewarming (MED)

Chinook The CH-470 rotary-wing, medium transport aircraft (AVN)

chlorine (CL) A toxic choking agent; short persistency, no decontamination required (NBC)

choking agent A chemical weapon that causes fluid buildup in the lungs and the consequent drowning effect (NBC)

chow Food (COLL)

CI Counterintelligence (INTEL)

CIA Central Intelligence Agency (INTEL)

CIB Combat Infantryman Badge (PERS)

CID Criminal Investigation Division (ADMIN)

CIF Central issue facility (LOG)

cifax Enciphered facsimile-machine signals (COMMS)

CINC Commander in chief (ADMIN)

cipher An alphanumeric element with a value or meaning other than its apparent one (COMMS)

cipher system Encrypted writing in which arbitrary symbols are substituted for plain text on a one-for-one basis (COMMS)

ciphony Enciphered speech communications (COMMS)

circular An official publication of general information of temporary value (ADMIN)

circular error (CE) A measurement of the accuracy of a map; the most commonly used circular error, 90, means that there is a 90 percent probability that a map feature falls within a specified range of its depicted position (NAV)

circulation control The restriction and monitoring of the movement of vehicles and personnel (TAC)

cirvis Communication instructions for reporting vital intelligence sightings (INTEL)

citation Written statement recounting the service or deed for which an award is made (PERS)

civil Nonmilitary, civilian (ADMIN)

civil affairs (CA) The relationship between military forces and civil authorities in an area under military control (ADMIN)

civil censorship Review and suppression, when necessary, of civilian communications entering and leaving areas under control of armed forces (CA)

civil defense Those passive measures taken to minimize the impact of hostile military action upon the civilian population (STRAT)

civil nuclear power A nation that has the technological ability to produce nuclear weapons but has decided not to develop them (STRAT)

civil twilight That time of incomplete darkness when the center of the sun is six degrees below the horizon (ADMIN)

civil works Activities relating to the conservation, development, and management of land and water resources (ENG)

civision Enciphered television signals (COMMS)

CK 1. Prosign meaning check group count (COMMS); 2. Symbol for the blood agent cyanogen chloride (NBC)

CL Symbol for the choking agent chlorine (NBC)

clandestine operation An intelligence or operational mission, the existence of which is concealed (INTEL)

clarity index A means of estimating the readability and effectiveness of text. The formula is as follows:

1. Count the number of sentences
2. Count the number of words
3. Divide the number of words by the number of sentences
4. Count the number of words with three syllables or more
5. Divide the number of these long words by the total number of words—target, 15 percent
6. Add the average sentence length to the percentage of long words
7. The sum is the clarity index. The target is 30; 20 or less is too abrupt, 40 or more is too difficult (ADMIN)

class A agent Commissioned or warrant officer authorized to disburse government funds in support of specified unit operations (ADMIN)

class B agent Commissioned or warrant officer authorized to disburse funds on a general basis under the authority of an accountable officer (ADMIN)

class B allotment An individual-authorized payroll deduction for monthly purchase of United States Savings Bonds (ADMIN)

class B1 allotment An individual-authorized payroll deduction for quarterly purchase of United States Savings Bonds (ADMIN)

class E allotment An individual-authorized payroll deduction for general assignment of pay to dependents, creditors, or investments (ADMIN)

classes of supply Categorization of materiel into groups in order to better identify and manage it; *See* Table A-10 (LOG)

classified information Official information that requires protection from disclosure in the interest of national security (INTEL)

class N allotment An individual-authorized payroll deduction for payment of premiums on life insurance (ADMIN)

Claymore mine Directional, antipersonnel fragmentation mine (WPN)

clear 1. To remove all ammunition from a weapon (WPN); 2. To authorize or approve (COLL)

clearing station A field medical facility providing emergency treatment for minor injuries and life support services pending evacuation (MED)

clock method Means of approximating a target's position relative to an observer and guns facing in the same direction: straight ahead is 12 o'clock; directly to the left, 9 o'clock; and right, 3 o'clock (TAC)

close air support Aircraft that deliver fire and bombardment against enemy positions that are sufficiently close to friendly positions to require careful coordination between air and ground assets (TAC)

close combat Small arms and hand-to-hand fighting (TAC)

closed Forces and their equipment in place and ready to perform their mission (TAC)

close defensive fires Fire on enemy command, observation, attack, and fire support positions intended to disorganize a forming attack (TAC)

close-in security Defense against very short range attack through employment of patrols, obstacles, and camouflage (TAC)

close interval The spacing of squad members in formation; base member places left hand on hip and soldier to his or her left moves to it (DC)

close march Command to assume a close interval while moving in formation (DC)

close-order drill Marching and synchronized movement, and manuals of arms performed by soldiers in close- and normal-interval formation (DC)

close ranks Preparatory command (command of execution: march) directing troops in open rank formation to return to a normal interval (DC)

close station A command that dismisses soldiers from a gun position (DC)

close support Fire support delivered on enemy units that are in contact with friendly units (TAC)

closure The time at which a unit and its equipment are fully deployed and ready to perform their mission (ADMIN)

cluster The basic unit of a minefield; may be either a live or an omitted cluster (TAC)

clutter Unwanted echoes from nontargets that appear as targets or obscuring images on a radar screen (INTEL)

CMD Command (ADMIN)

CMF Career management field (ADMIN)

CN solutions Symbol for a series of tearing agents (NBC)

CO **1.** Commanding officer (ADMIN); **2.** Company (ADMIN)

coarse setting The preliminary use of the main scale of a sight in the laying of a gun (ARTY)

COAX Coaxial machine gun (WPN)

coaxial machine gun A large-caliber, belt-fed automatic gun mounted on a tank's turret, parallel to the main gun (ARM)

COB Close of business—the end of the duty day (ADMIN)

Cobra The AH-15 light, rotary-wing attack aircraft (AVN)

Cocked Pistol Exercise term for DEFCON 1, a state of alert (STRAT)

code A symbol or group of symbols that represents units of plain text of any length (COMMS)

code of conduct Positive mission guidance for soldiers who become prisoners of the enemy; *See* Appendix B (PERS)

code word A classified word or combination of words used to identify a classified operation plan or system (INTEL)

codress A message address that is embedded within the encrypted text (COMMS)

COHORT Cohesion, Operational Readiness Training (TNG)

COIN Counterinsurgency (UW)

COL Colonel (PERS)

COLA Cost-of-living allowance (ADMIN)

cold dry clothing Cold wet clothing items plus additional insulating items sufficient to protect against temperatures as low as minus 50 degrees Fahrenheit (PERS)

cold war Conflict between nations by means other than all-out war, i.e., espionage, diplomatic conflict, covert operations, and propaganda (STRAT)

cold wet clothing Water repellent and insulating items of clothing sufficient to protect against temperatures above 14 degrees Fahrenheit (PERS)

collateral damage Unintended destruction to persons or property other than the target of bombing (WPN)

collation The assembly of information into meaningful, unduplicated categories; part of the processing step of the intelligence cycle (INTEL)

collection The gathering of raw information and its delivery to the appropriate processing unit (INTEL)

collective protection Sheltered space in which the air is filtered, permitting the individual occupants to relax their MOPP status (NBC)

collocation The placement of two or more units at the same or adjacent sites to enhance their ability to interact efficiently (ADMIN)

colonel An officer; pay grade O-6 (PERS)

color guard Detachment of four to six enlisted soldiers whose ceremonial duty is to bear the colors (DC)

colors **1.** The National Color: the U.S. flag (DC); **2.** An organizational flag (DC)

COLT Combat observation/lasing teams (ARTY)

column A line formation whose elements are one behind the other (DC)

column half left (right) Preparatory portion of a two-part command directing a column to execute a 45-degree turn (DC)

column left (right) Preparatory portion of a two-part command directing a column to execute a 90-degree turn (DC)

combatant Individual soldier of belligerent forces subject to the laws, rights, and duties of war (LAW)

combat arms Those branches of the Army whose members directly participate in battle—aviation, infantry, field artillery, air defense artillery, armor, and engineers (ADMIN)

combat commander's insignia Green cloth tab worn on the epaulet of enlisted personnel or officers in the combat chain of command (PERS)

combat control team Personnel capable of establishing landing, navigational, and communications facilities in support of airborne operations (TAC)

combat engineer vehicle Fully tracked and armored vehicle, equipped with a 165 mm demolition gun, boom and winch, dozer blade, and other machine guns (ENG)

combat information Data on the enemy's tactical deployment that is too perishable to be evaluated and is therefore passed on to the combat commander prior to its processing into intelligence (INTEL)

combat intelligence Analyzed data on enemy, weather, and terrain that has an impact on the immediate tactical situation (INTEL)

combat loading The stowage for transport of equipment and personnel in a manner compatible with their rapid tactical deployment on arrival (LOG)

combat multiplier Any factor that enhances the efficient employment of maneuver assets rather than only adding to them (MS)

combat power The destructive force that a unit or formation can generate against the enemy (MS)

combat reconnaissance Observation of enemy elements in contact with friendly forces, either before or during an operation (TAC)

combat service support Those administrative functions not related to combat operations or their direct support, such as finance, health services, supply and food service, chaplain services, and transportation (ADMIN)

combat support Operational assistance to combat units (ADMIN)

combat support hospital A mobile medical treatment facility (MED)

combat tire Tire designed to run while flat for a limited distance in an emergency (LOG)

combat trains Supply distribution elements that support maneuver units with ammunition, fuel, and maintenance (LOG)

combination firing circuit Independent parallel systems for detonating a set of charges, either of which is capable of exploding the circuit (WPN)

combination vehicle A towing vehicle and its towed load (LOG)

combined arms The coordinated use of more than one combat branch during operations (TAC)

combined command A directive that is executed without a preparatory command (DC)

combined force A military element containing units from two or more nations (STRAT)

combined intelligence Intelligence shared among the United States and its allies in support of joint operations (INTEL)

combined training Exercises in which units that would deploy together in wartime train together (TNG)

come-as-you-are war A defense emergency of sufficient urgency to require deployment of Reserve units without time for further training additional personnel and equipment (COLL)

COMINT Communications intelligence (INTEL)

COMJAM Communications jamming (EW)

command **1.** The authority to direct, and responsibility for, the mission and welfare of military forces (MS); **2.** A lawful directive (MS)

command and control The exercise of authority, by a duly designated officer, over assigned forces in accomplishment of a mission (MS)

commander Officer responsible for a unit's training and operation (MS)

commander of the guard Commissioned or noncommissioned officer responsible for the instruction, discipline, and duty performance of a guard detail (ADMIN)

Commander's Call Formal, weekly instruction of unit members to meet the requirements of the Command Information Program (TNG)

command group Those officers key to the operation of a forward echelon (ADMIN)

command information Information provided to soldiers by their commanders to help them understand their role within their unit and the Army (ADMIN)

command net A communications network connecting a command echelon with its subordinate commanders (COMMS)

commando Soldier trained to conduct raids inside enemy territory (UW)

command of execution Completion of a two-part command, the issuance of which conveys that it should be performed (DC)

command post (CP) The location of a unit's headquarters commander and his or her staff (TAC)

command post exercise (CPX) A medium-cost simulation of maneuvers designed to train and test command, control, and communications by staff elements (TNG)

command sergeant major (CSM) A senior enlisted noncommissioned advisor to a commander at a battalion level and higher; pay grade E-9 (PERS)

commercial Nontactical; equipment that is essentially the same as the civilian version (LOG)

commercial air movement number A control number assigned to groups of 15 or more soldiers traveling within CONUS by commercial aircraft (ADMIN)

commissary A grocery store (LOG)

commission A written appointment granting rank and authority as an officer (ADMIN)

commissioned officer A soldier whose authority to act within his or her duties on behalf of the United States derives from presidential designation (PERS)

common table of allowances An authorization document for the distribution of certain ordinary supplies based on type of organization or an individual's duties (LOG)

common use facilities DOD facilities that are used by more than one service (LOG)

common user airlift service Military transport services open to all DOD branches (LOG)

communications intelligence (COMINT) Information of tactical and strategic value that has been derived from intercepted communications of foreign nations (INTEL)

communications security (COMSEC) The full range of measures used to deny access to friendly signal transmissions; includes cryptosecurity, transmission security, emission security, and physical security (EW)

communications terminal Point in a network where messages can be transmitted or received (COMMS)

communications zone That rear area of a theater of operations that includes lines of communication, supply and evacuation routes, and other support functions (TAC)

commuted ration Monetary payment in lieu of subsistence to enlisted personnel authorized to mess separately (PERS)

COMMZ Communications zone (TAC)

company A unit at a command level below a battalion and above a platoon; equivalent to a battery or troop (ADMIN)

company grade Officers who serve at company level—lieutenants and captains (ADMIN)

compartmentation The segregation of various types of intelligence in order to restrict its dissemination to those with a "need to know" (INTEL)

compartment of terrain A sector bounded on at least two sides by vegetation, mountains, buildings, or other obstructions that limit the ability to see in and out of it (TAC)

COMPASS Computerized Movement Planning and Status System (LOG)

compass compensation The adjustment of a compass reading to allow for the effect on it of nearby metal objects (NAV)

compass north The uncorrected direction indicated by a magnetized compass needle (NAV)

compass rose A graphic symbol that indicates a map's directional orientation (NAV)

complementary angle of site An aiming correction that compensates for a nonrigid trajectory (ARTY)

complementary range Meters of range added to compensate for the complementary angle of site (ARTY)

complete round Ammunition containing all the components necessary to fire it—a projectile, propellent, and possibly a fuse, primer, and cartridge case (ORD)

complex working Radio communication via transmission on one frequency and reception on another (COMMS)

component end item A specialized machine or system that is a subsystem of a larger assembly often held at a higher echelon; a subunit of a basic end item (LOG)

compound helicopter A rotary-wing aircraft with an auxiliary propulsion system to boost forward speed (AVN)

compromise The actual or potential exposure of classified information to unauthorized personnel (INTEL)

computed air release point An airdrop position based on calculation of wind, airspeed, and altitude rather than observation of a trial drop (ABN)

COMSEC Communications security (INTEL)

COMSEC accounting A system that provides close control of and responsibility for classified communications security devices and material (COMMS)

concealment The shielding from observation (TAC)

concentrated fire Fire from two or more weapons directed toward a single target area (TAC)

concept of operations A brief statement describing, in general terms, how a mission is to be accomplished (TAC)

concurrent jurisdiction Area where responsibility for law enforcement is shared by the state and federal governments (LAW)

condition code An alphabetic character indicating the serviceability of materiel for use or issue (LOG)

cone of fire The pattern that results from a burst of fire, due to gun vibration, wind, and variations in ammunition (WPN)

CONEX Container express (LOG)

confidence course A physically challenging obstacle course designed to bolster a soldier's own estimate of his or her ability (TNG)

confidential A classification for national security material requiring protection, the unauthorized disclosure of which could cause damage to national security (INTEL)

confusion agent An intelligence operative whose objective is to disinform the enemy's intelligence or counterintelligence apparatus (INTEL)

confusion reflector Any device that disrupts radar signals, such as chaff or corner reflectors (TAC)

connection survey A correlation of target area and position area surveys (ARTY)

CONOPS Continuous operations (ADMIN)

consignee The receiver of cargo as indicated on a shipping document (LOG)

consignor The sender of cargo as indicated on a shipping document (LOG)

consolidated A support activity serving multiple units or all units within an area facility (ADMIN)

consolidation of position Development of a newly acquired area to enhance its offensive or defensive potential (TAC)

consolidation phase of military government The period that begins with establishment of control by ground forces and continues through hand over to the occupation force (CA)

consumable supplies An accounting classification that includes expendable supplies and nonexpendable supplies valued at less than $200 (LOG)

contact **1.** Engagement with the enemy (TAC); **2.** Establishment of radio communication (COMMS)

contact patrol Mobile detachment with the mission of continuously harassing and/or reconning nearby enemy units (TAC)

contact point An easily identifiable point on the terrain where communication is required between a maneuvering unit and a control element (TAC)

contain To prevent the maneuver or withdrawal of enemy forces (TAC)

contamination The deposit or absorption of a chemical, biological, or radioactive agent capable of harming personnel (NBC)

contingency operations Politically sensitive peacekeeping operations characterized by short, rapid protection of joint or combined forces in conditions short of war (STRAT)

contingency requisitions Materiel requests that are submitted in anticipation of approval of a particular operation plan (LOG)

contingency support stocks That portion of the general war reserves dedicated to the initial resupply of CONUS forces deployed in contingency operations (LOG)

continuity of command The concept that the direction and control of military forces supersedes in importance any transition between individual commanders (MS)

continuous fire Delivery of rounds at a constant, normal rate, without adjustment or correction (ARTY)

continuous wave Transmission of unmodulated radio waves; suitable for broadcast of Morse code tones (COMMS)

contour flying Low-altitude aviation operations that follow the ground terrain to minimize the vulnerability of aircraft to ground observation and/or fire (AVN)

contour interval An indication of the terrain's slope; the difference in elevation between adjacent contour lines (NAV)

contour line A continuous line on a map connecting points of equal elevation (NAV)

control group Administrative grouping for those Reservists not assigned to troop program units or the staff and faculty of a U.S. Army Reserve school (ADMIN)

controlled firing area Weapons range with restrictions to eliminate hazards to aircraft in flight (TNG)

controlled information Data intentionally communicated to the enemy in an attempt to manipulate their actions (INTEL)

controlled inventory items Items of materiel requiring special accounting or handling to safeguard them, including sensitive items, classified items, and items prone to pilfering (LOG)

controlled mine A mine that can be made safe and reactivated by the user after it is laid (ORD)

controlled net Communications circuit in which outstations may transmit only with permission from the control station (COMMS)

controlled pattern Air delivery of materiel connected by webbing or rope to prevent its excessive dispersal (ABN)

controlling authority The command responsible for the establishment and operation of a cryptonet (COMMS)

control point A station along a route from which soldiers provide information and regulate traffic flow (TAC)

control station (CT) The radio facility that directs a communication network (COMMS)

CONUS Continental United States; a unit or activity within the contiguous 48 states (ADMIN)

conventional weapon A weapon that is not nuclear, biological, or chemical (WPN)

convergence The decreasing distance between meridians as one travels from the equator to either pole (NAV)

convoy A group of vehicles organized to travel together (LOG)

cook off The detonation within a weapon due to excessive heat (WPN)

COOP Continuity of operation plan (ADMIN)

coordinated fire line A line beyond which artillery fire may be delivered without the specific approval of the commander responsible for the area (TAC)

coordinates Horizontal and vertical lines on a map used to pinpoint a position (NAV)

coordinating point Site along a route of movement at which the maneuvering element must communicate with any controlling element (TAC)

coordination Communication intended to synchronize activity between elements (ADMIN)

Copperhead The M712, a laser-guided, antiarmor artillery round (ORD)

cordon A rank of soldiers used to honor passing dignitaries (DC)

corner reflector A metallic device designed to reflect radar signals, either to serve as a marker for friendly systems or as a false target to the enemy (INTEL)

corporal An enlisted, junior noncommissioned officer in the pay grade E-4 (PERS)

corps A unit larger than a division and smaller than an army, normally consisting of two or more divisions plus auxiliary units (ADMIN)

corrected range Distance to the target, adjusted for weather, ammunition, gun wear, or any other variation (ARTY)

corridor A long, narrow compartment of terrain parallel to a force's direction of movement (TAC)

COSCOM Corps support command (ADMIN)

counter Prefix meaning against or to oppose (MS)

counterattack Offensive action by defending forces against attacking forces, with the general objective of preventing the attacking forces from attaining their objective (TAC)

counterbattery fire Fire delivered against the enemy's indirect fire positions (ARTY)

counterespionage Counterintelligence targeted specifically to detect and neutralize hostile espionage activities through identification, penetration, and/or repression (INTEL)

counterfire Fire directed against enemy weapons (ARTY)

counterinsurgency The spectrum of measures taken by a government to defeat subversive elements—military, civic action, psychological operations, or economic (UW)

counterintelligence Activities concerned with identifying and thwarting hostile intelligence services or agents engaged in espionage, sabotage, or subversion (INTEL)

counterintelligence technical survey Sweeping of an area to ensure the absence of remote listening devices or other sensors (INTEL)

countermobility Defensive preparation of friendly territory to slow or prevent the enemy's advance (TAC)

counteroffensive An attack in response to an enemy attack meant to regain the initiative by putting the enemy back in a defensive mode (TAC)

counterpreparation fire Intensive prearranged fire delivered against enemy troops forming for an imminent attack (ARTY)

counterrecoil The forward movement of a gun to its original position after the recoil of firing (ARTY)

countersign A challenge and password; code words used to ascertain whether an unknown individual is friendly or enemy (TAC)

countersurveillance Passive and active measures taken to prevent observation by the enemy (INTEL)

count off Two-part command directing members of a formation to call out, in sequence, their numerical positions in the formation (DC)

country team Members of a U.S. diplomatic mission in support of a foreign host nation (STRAT)

courier A means of secure communication—a soldier who carries documents or other material from one place to another (COMMS)

court-martial Any of several types of military tribunals authorized to try cases of violation of the Uniform Code of Military Justice (LAW)

court of inquiry A board of three or more officers appointed to hear accusations against military personnel (LAW)

court of military appeals A court of civilian judges empowered to hear appeals of military courts-martial (LAW)

cover 1. To align the elements of a formation, from front to rear (DC); 2. A fire control order directing a battery to prepare to engage a target (AD)

cover code The portion of a target that is hidden by terrain from observation and fire (TAC)

covered movement Maneuver of troops that is protected by delivery of fire to suppress enemy interference (TAC)

covering fire Fire delivered in support of troops within the range of enemy small arms fire (TAC)

covering force Any element whose mission is to intercept, engage, delay, disorganize, or deceive the enemy before it can attack a friendly main body (TAC)

covert operations Intelligence or special operation whose sponsor is hidden, permitting plausible denial for its responsibility (UW)

CP Command post (TAC)

CPE Collective protection equipment (NBC)

CPL Corporal (PERS)

CPO Civilian personnel office (ADMIN)

CPR Cardiopulmonary resuscitation—restoration of heartbeat and breathing (MED)

CPT Captain (PERS)

CPX Command post exercise (TNG)

CQ Charge of quarters (ADMIN)

crater analysis Determination of the source and caliber of artillery fire by examination of the round's impact point (INTEL)

cratering charge An explosive that is buried beneath a road or runway at a depth designed to blow large pits in it (ORD)

CRD Criminal Records Directorate (ADMIN)

credit system of supply The allotment of materiel to units on a monetary or per-item basis—materiel is furnished on request and charged against the unit account (LOG)

creeping barrage Fire in which all guns maintain their relative position while shifting their range of fire in only small increments (ARTY)

crest A terrain feature of such altitude that it shields targets behind it from direct fire; such targets are said to be crested (TAC)

crest clearance Elevation of a gun to a point where rounds will not impact obstacles between the muzzle and target (ARTY)

CRG **1.** Counterfire reference grid (ARTY); **2.** Communications relay group (COMMS)

CRITIC Critical intelligence (INTEL)

critical facility Site of such importance to national defense that it is a likely enemy target (TAC)

critical intelligence Information that is required by a commander in order to support a timely decision in response to actual or potential enemy actions (INTEL)

criticality The degree to which an asset is essential to mission accomplishment (AD)

critical node A tactical position or link in a communications system, the destruction of which immediately degrades the combat effectiveness of a force (TAC)

crossing area A number of crossing sites under the control of a single commander (TAC)

cross talk The spillover or bleeding of one communications circuit into another (COMMS)

cross tell The exchange of information between units at the same operational level (ADMIN)

cross training The exposure of soldiers to tasks and specialties closely related to their own, for familiarization and increased flexibility (TNG)

cruise missile A guided projectile that can be programmed to follow specified routes, adjusting its flight path based on its own interpretation of its current position (WPN)

cryptanalysis The process of converting encrypted text to plain text without knowledge of the key employed (INTEL)

crypto Designation for classified keying material related to secure communications (COMMS)

cryptodate The date that determines the cryptokey to be used (COMMS)

cryptodevice A device used to simplify encryption or decryption but containing no cryptoprinciple (COMMS)

cryptoequipment Any device that employs a cryptoprinciple (COMMS)

cryptography The science of efficiently converting plain text into securely unintelligible text (COMMS)

cryptology The science and study of secret communications (COMMS)

cryptomaterial Any devices, documents, or equipment essential to the encryption or decryption of secret communications (COMMS)

cryptonet Two or more subscribers to a common cryptosystem (COMMS)

cryptoperiod A time span during which certain cryptovariables are to be in effect (COMMS)

cryptoprinciple The logical steps by which a message is converted to unintelligible form and back to usable form (COMMS)

cryptoservice A secure message regarding difficulties or irregularities in the encryption or decryption of other messages (COMMS)

CS **1.** Critically sensitive (INTEL); **2.** Symbol for the tearing agent o-chlorobenzalmalononitrile (NBC)

CSA Clean shelter area (NBC)

CSAR Combat search and rescue (AVN)

CSM Command sergeant major (PERS)

CSR Combat stress reaction (MED)

CSS Combat service support (ADMIN)

CSTVRP Computer Security Technical Vulnerability Reporting Program (ADMIN)

CTA Common table of allowances (LOG)

CTC Commanders' target criteria (ARTY)

CTT Common task training (TNG)

CTX Combined training exercise (TNG)

CUCV Commercial utility/cargo vehicle (LOG)

cueing The use of a collection asset to alert another system to probe for additional or confirming information (INTEL)

cupola A dome-shaped, armored gun mounting or commander's hatch on top of a tank's turret (ARM)

curfew Period (normally overnight) during which the public is forbidden to be on the streets (CA)

current files Active records for the current year or ongoing operations (ADMIN)

cut The point of intersection of two direction-finding lines of bearing (EW)

cutout An individual or other intermediary used to prevent direct contact between and knowledge of clandestine elements (INTEL)

CVC Combat vehicle crew (ARM)

CW Continuous wave (COMMS)

CWIE Container, weapon, individual equipment (ABN)

CWO Chief warrant officer (PERS)

CX **1.** Complex working (COMMS); **2.** Symbol for the blister agent phosgene oxime (NBC)

CY **1.** Calendar year (ADMIN); **2.** Copy (ADMIN)

cyanogen chloride (CK) Toxic blood agent—no decontamination necessary in the field (NBC)

cyclic rate of fire The maximum number of rounds that can be fired per minute (WPN)

CZ Combat zone (TAC)

D

DA **1.** Department of the Army (ADMIN); **2.** Symbol for the vomiting agent diphenylchloroarsine (NBC)

DACO Departure airfield control officer (ABN)

DAG Division artillery group (OPFOR)

DAME Division airspace management element (AD)

D & C Drill and ceremonies (TNG)

danger Advice in a call for fire that friendly forces are within 600 to 1,500 meters of the target (ARTY)

danger close Advice in a call for fire that friendly forces are within 600 meters of the target (ARTY)

danger space The area between a weapon and its target where the round's trajectory does not rise above the height of the average standing soldier (1.8 meters) (WPN)

DANTES Defense activity for nontraditional education support (TNG)

DAO Division ammunition officer (PERS)

data link A means of transmitting information from one location to another (COMMS)

date break Date on which COMSEC keys and procedures change (COMMS)

date of rank (DOR) Effective date of last promotion (ADMIN)

date-time group A six-digit time designation—the first two digits represent the date, the second two the hour, and the final two the minutes (ADMIN)

daylight traffic line Point beyond which vehicles are not permitted during daylight (ADMIN)

day room Lounge in a barracks (COLL)

DBDU Desert battle dress uniform (LOG)

DC Symbol for the vomiting agent diphenylcyanoarsine (NBC)

DCA Defensive counterair (AVN)

DCI Director, Central Intelligence (INTEL)

DCO Deputy commanding officer (ADMIN)

DCS Deputy chief of staff (ADMIN)

D-day The date on which an operation is to commence; a planning point within the wartime manpower planning system (ADMIN)

DE Prosign meaning this is (COMMS)

deactivate To render a mine or other explosive device safe or inert (ORD)

deadlined Equipment that is inoperable (LOG)

dead reckoning Estimate of one's position based on speed, direction, and elapsed time since last known position (NAV)

dead space Area within a gun's range that cannot be hit by that weapon (WPN)

debarkation The unloading of a ship or aircraft (LOG)

debriefing The assisted review and extraction of information from a friendly asset (INTEL)

deception Any measure designed to mislead an enemy into actions counterproductive to his or her interests (TAC)

decisive terrain Position(s) that must be held to control the outcome of an engagement (TAC)

DECL Declassify (INTEL)

declassification The determination by proper authority that restriction of access to certain information is not in the national interest (INTEL)

declination Position on the celestial sphere (in the sky) measured in angular distance from the celestial equator (NAV)

decontamination The neutralization or removal of nuclear, chemical, or biological agents (NBC)

decoration A distinctively designed medal awarded in recognition of heroism or meritorious service or achievement (PERS)

decrypt To convert from encrypted text to plain text through use of a cryptosystem (COMMS)

dedicated Committed to a particular role or mission (COLL)

deep battle Offensive action that engages not only the enemy's readily accessible front echelons but also successive rear echelons through employment of airmobile and/or airborne assets and flanking maneuvers (STRAT)

deep fording The crossing of a water obstacle by a vehicle with its wheels in contact with the ground, when special waterproofing equipment is required (TAC)

deep supporting fire Fire directed at support and reserve echelons instead of at enemy forces in contact (ARTY)

DEERS Defense Enrollment Eligibility Reporting Systems (ADMIN)

DEFCON Defense readiness condition (ADMIN)

defector A person who repudiates his or her native country in favor of an opposing nation (LAW)

defense area The area extending from the forward edge of a battle area to a command's rear boundary (TAC)

Defense Distinguished Service Medal Decoration made in recognition of exceptionally meritorious service while in a joint service position of unique, great responsibility (PERS)

defense in depth A system of mutually supporting positions designed to absorb, delay, and weaken an attack (STRAT)

defense information Data that, while not classified, are not common knowledge and therefore are of potential value to an enemy (INTEL)

defense in place Resistance to an enemy offense without retreat, rather than by relocating to successive positions (TAC)

Defense Meritorious Service Medal Decoration made in recognition of meritorious, noncombat joint service (PERS)

defense readiness conditions (DEFCON) The system of military alert postures designed to mirror the level of threat faced (STRAT)

Defense Superior Service Medal Decoration made in recognition of meritorious service while in a joint position of significant responsibility (PERS)

defensive fire Fire delivered in support of a unit engaged in a defensive action (TAC)

defensive position A fortified area from which to deliver fire (TAC)

defensive zone A band of terrain parallel to the front that contains two or more organized battle positions (TAC)

defilade A firing position that affords protection from observation and direct fire (TAC)

deflection The setting of the line of fire to a desired direction (WPN)

deflection error Distance to the left or right between a target and the impact point of a round or mean impact point of a salvo (ARTY)

defoliating agent A chemical that causes plant life to shed its leaves (NBC)

DEFREP Defense readiness posture (ADMIN)

degauss To demagnetize or erase any magnetic information storage material, e.g., tape or disk (COMMS)

delay fuze Detonation initiator designed to permit its projectile to penetrate a target before exploding (ORD)

delaying action A retrograde operation in which space is sacrificed for increased time to punish the enemy without decisive engagement (TAC)

delegation of authority The assignment by a commander to a subordinate of his or her power to administer a specifically limited function (ADMIN)

deliberate Any operation conducted thoroughly while not in contact with the enemy (TAC)

deliberate fire Fire delivered at a rate that permits adjustment or conserves ammunition (ARTY)

delivery error The average inaccuracy of a weapons system (WPN)

demand satisfaction The immediate filling of 90 percent of valid, authorized materiel requisitions (LOG)

demodulation The process of extracting information from a transmitted carrier radio wave (COMMS)

demolition The intentional destruction of structures or materiel by any means (TAC)

demolition guard A detachment positioned to defend a specified point target, such as a bridgehead, until it is no longer useful to friendly forces (TAC)

demonstration A show of force intended to deceive rather than achieve a decisive tactical objective (TAC)

denial measure The prevention of the enemy from benefiting from objects or positions of tactical value (TAC)

Department of Defense (DOD) The executive agency, headed by the Secretary of Defense, that includes the War Council and the Organization of the Joint Chiefs of Staff, and through them administers the Departments of the Army, Navy, and Air Force (ADMIN)

Department of Defense identification code A four-character code that identifies and specifies the interchangeability of ammunition (ORD)

dependency and indemnity compensation Veterans Administration payment to dependent survivors of soldiers who die in the line of duty (ADMIN)

DEPEX Deployment exercise (TNG)

deployment The movement of forces to desired operations areas (STRAT)

depot A large-scale combat service support facility (LOG)

depth The extension of operations in space, time, and resources (MS)

deserter A soldier absent without authority in excess of 30 days and who has been dropped from his unit's rolls (LAW)

designation of days and hours A series of coded planning points for mobilization and deployment within the wartime manpower planning system (ADMIN)

destroyed A target that cannot function as it was intended or be restored to a usable condition (TAC)

destructor A device designed to destroy a friendly missile or vehicle for safety reasons or to prevent its compromise (ORD)

detachment A subunit of a larger unit, specifically tailored for a given mission (ADMIN)

detail A work party or very-short-term assignment (ADMIN)

detainee Prisoner (LAW)

detainer A notice to civil authorities that a soldier in their custody is also wanted by the Army (LAW)

deteriorating supplies Items expected to become unusable during one to two years of storage (LOG)

deterrence The prevention of an enemy action by the existence of a credible threat of unacceptable counteraction (STRAT)

detonating charge An explosive used to set off a larger, high explosive charge (ORD)

detonating cord A high explosive in a flexible fabric tube used to transmit explosive shock from one charge to another; commonly call a det (ORD)

deuce-and-a-half A 2.5-ton cargo truck (COLL)

devastation Destruction beyond that necessary to secure a military objective, especially after surrender of the area (LAW)

DEW Directed energy weapon (WPN)

DF **1.** Direction finding (EW); **2.** Deposition form (ADMIN)

DFR Dropped from rolls (ADMIN)

DIA Defense Intelligence Agency (ADMIN)

DIAM Defense Intelligence Agency Manual (INTEL)

Diamond Lil The M157 projected charge demolition kit used to clear a four-meter-wide path through an antitank minefield (TAC)

diaphosgene (DP) A toxic choking agent; decontamination not necessary in the open (NBC)

DIC **1.** Dependency and indemnity compensation (ADMIN); **2.** Document identifier code (LOG)

dih dah A Morse code operator (COLL)

dining in A formal banquet gathering of a unit's members (COLL)

dining out A formal banquet gathering of a unit's members and spouses (COLL)

direct action mission A special operations mission of specific, limited objectives, normally conducted in a hostile area (SF)

direct communication Message traffic authorized to skip normal intermediate echelons (COMMS)

directed net A communications circuit in which a net control station must authorize all messages (COMMS)

direct exchange The issue of equipment or supplies in exchange for a turn in of identical, unserviceable items (LOG)

direct fire Gunfire directed at a target visible to the aimer (TAC)

direction finding The use of direction-sensitive receivers to determine the origin of enemy radio signals (INTEL)

direct laying The positioning of a gun with its sights aligned on the target (ARTY)

direct plotting A single determination of range and azimuth data to enable an entire battery to fire on a moving target (ARTY)

direct pressure All offensive action other than encirclement against a retreating enemy, in order to deny them an opportunity to reorganize (TAC)

directrix The central axis of a gun's field of fire (ARTY)

direct support Command relationship in which a dedicated supporting unit performs missions on priority call from another unit within the command (LOG)

direct support maintenance Repair, overhaul, and modification of equipment by units organized to support other units by maintaining their equipment and returning it directly to them (LOG)

DIS Defense Investigative Service (INTEL)

disaffected person An individual who lacks loyalty to his or her government (CA)

discharge without honor Characterization of a soldier's term of service as less than honest and faithful, when the soldier's character has been rated fair or poor and when a dishonorable discharge is not applicable (ADMIN)

DISCOM Division support command (LOG)

dishonorable discharge Characterization of a soldier's term of service upon conviction and sentencing by a general court-martial (ADMIN)

disinformation Information disseminated to deceive or mislead the enemy (INTEL)

dismounted On foot; not in a vehicle (INF)

dispatch route A road over which each vehicle must have specific authority for movement (ADMIN)

dispensary An issue point for medicine and medical supplies (MED)

dispersal The increased spacing between forces to minimize the potential damage from a single enemy round (TAC)

dispersion **1.** The scattering of the impact points of rounds fired under identical conditions (ARTY); **2.** The spreading by wind of airborne radioactive fallout or chemical agents (NBC); **3.** The normal distribution of air-delivered parachutists or bundles resulting from wind and aircraft movement (ABN)

dispersion error The distance between the point of impact of a round and the mean point of impact of a group of rounds (WPN)

dispersion pattern The normal distribution of impact of a series of rounds fired under conditions as nearly identical as possible (ARTY)

dispersion rectangle A table depicting the distribution of rounds fired based on the same firing data (ARTY)

displace To move from one position to another (TAC)

displaced person A refugee; a civilian forced by conflict from his or her native land (LAW)

disposition The array or status of forces (MS)

disposition form (DF) A blank sheet of paper, formatted for expedient, informal written messages (ADMIN)

dissemination The timely distribution of finished intelligence to appropriate users with a need to know it (INTEL)

Distinguished Flying Cross Decoration made in recognition of heroism or extraordinary achievement while in aerial flight (PERS)

Distinguished Service Cross Decoration made for extraordinary heroism in military operations against an enemy of the United States (PERS)

Distinguished Service Medal Decoration made for meritorious service not involving personal bravery in a duty of great responsibility (PERS)

Distinguished Unit Citation (DUC) Decoration made to a unit as a whole, representing a degree of heroism equivalent to that required for award of the Distinguished Service Cross to an individual (ADMIN)

DISUM Daily intelligence summary (INTEL)

DIV Division (ADMIN)

DIVAD Division air defense (ADMIN)

DIVARTY Division artillery (ADMIN)

diversion Any activity intended to mislead the enemy by drawing their attention away from meaningful friendly activity (MS)

diversity reception The use of multiple antennas and/or frequencies to minimize signal fading (COMMS)

division A tactical unit at a command level below a corps and above a brigade (ADMIN)

division support command (DISCOM) A unit within a division with the specific mission of managing the supply, transportation, maintenance, and medical support for the division (ADMIN)

DLA Defense Logistics Agency (ADMIN)

DLAT Defense Language Aptitude Test (TNG)

DLI Defense Language Institute (TNG)

DLIC Detachments left in contact (TAC)

DLPT Defense Language Proficiency Test (TNG)

DM Symbol for the vomiting agent adamsite (NBC)

DMD Digital message device (COMMS)

DME Distance measuring equipment (ARTY)

DMG Distinguished military graduate (TNG)

DMZ Demilitarized zone (ADMIN)

DOB Date of birth (ADMIN)

doctrine Normal conduct of operations; standard policies that are authoritative but require judgment in their application (MS)

document number A 14-digit control number assigned at unit level to identify a transaction affecting the inventory balance of a supply item (LOG)

document register A log used to list and control records of supply transactions (LOG)

DOD Department of Defense (ADMIN)

DODAAC Department of Defense activity address code (ADMIN)

DODAC Department of Defense ammunition code

DODIC Department of Defense identification code (ADMIN)

dog tag Personal identification tag (COLL)

DOI Date of introduction (LOG)

dominant user concept The principle that states that the service (e.g., Army, Navy, or Air Force) that is the primary consumer of a service or system should manage the support work load for it (LOG)

door bundle A supply package delivered by airdrop; can be free dropped or parachute delivered (ABN)

DOR Date of rank (ADMIN)

DOS Days of supply (LOG)

DOSAAF Soviet paramilitary organization for young people (OPFOR)

dose rate contour line A line on a map or overlay connecting points of equal radiation contamination (NBC)

dosimetry The measurement of exposure to radiation (NBC)

double action Characteristic of small arms in which a single trigger pull both cocks the hammer and fires the weapon (WPN)

double agent An intelligence officer who is assigned to work for one nation but actually provides intelligence for, and is loyal to, another nation (INTEL)

double apron fence An obstacle fence with barbed wire on both sides (TAC)

double canopy forest Wooded area with two distinct levels of treetops (TAC)

double envelopment To surround from three directions; employment of three offensive elements of engagement—normally the front, plus both flanks or either flank and the rear (TAC)

double lane A cleared path through a minefield wide enough (15 meters) to permit two-way vehicular traffic (TAC)

Double Take Exercise term for DEFCON 4, a state of alert (STRAT)

double time Movement of troops in formation at the rate of 180 paces per minute (DC)

downed aviator point A ground rally point to which fliers should proceed for recovery (AVN)

down link A channel between an airborne platform and a ground-based radio station (COMMS)

down range Toward the enemy, targets, or combat (COLL)

down time **1.** Period during which equipment is not mission capable for maintenance reasons (MAINT); **2.** Interval between requisition and receipt of materiel (LOG)

DP Symbol for the choking agent diaphosgene (NBC)

DPA Data processing activity (ADMIN)

DPICM Dual-purpose improved conventional munitions (ORD)

Dragon The M47 man-portable, antitank/assault missile (WPN)

draw A stream course that has not developed a valley floor (NAV)

dream sheet Form on which a soldier requests his or her preferred duty assignments (COLL)

dress To align (COLL)

dress left/right Preparatory command (command of execution—dress) directing soldiers in formation to align themselves on the soldier to the indicated side (DC)

drill Standardized, precision movement of individuals and soldiers in formation for the purpose of developing discipline and coordination of effort (DC)

drill sergeant A noncommissioned officer trained to conduct the administration of basic combat or advanced individual training to new soldiers (TNG)

drone An unmanned, remotely piloted vehicle (LOG)

droop stop Supports placed under a helicopter's rotors while it is on the ground (AVN)

drop **1.** Direction by a fire observer to decrease range (ARTY); **2.** A deployment mission (ABN)

drop zone (DZ) Area identified and marked for parachute landings (ABN)

dry gap bridge A bridge used to span obstacles (e.g., ditches and craters) that do not contain water (LOG)

DS Direct support (ADMIN)

DS 2 Decontaminating solution #2: a ready-to-use solution that is effective against all known toxic chemical and biological agents except bacterial spores; irritating to eyes and skin if contact is prolonged beyond 30 minutes (NBC)

DSA Division support area (ADMIN)

DSB Direct support battery (ARTY)

DSM Distinguished Service Medal (PERS)

DSSA Direct Supply Support Activity (LOG)

DSSCS Defense Special Security Communications System (COMMS)

DSU Direct support unit (LOG)

DTG Date time group (ADMIN)

D to P concept A system for managing the flow of materiel in support of operations (LOG)

dual capable unit A unit capable of executing both conventional and nuclear missions (ADMIN)

dual firing circuit An assembly of charges that can be detonated by either of two independent systems (WPN)

dud A munition that fails to explode as planned (ORD)

dud probability The expected percentage of misfires among a number of firings (ORD)

due in Requisitioned materiel for which a scheduled delivery date has been received (LOG)

due out Inventory that is on hand but committed for issue (LOG)

dummy traffic A message consisting of meaningless characters, sent as a channel test or to occupy enemy resources (COMMS)

dump A temporary storage area, normally in the open (LOG)

durable materiel A subclass of the nonexpendable materiel category that includes items, such as tools, that tend to wear out or be damaged and therefore are accounted for differently than expendable materiel (LOG)

dustoff A medical evacuation helicopter (AVN)

duty officer Soldier detailed to serve as a unit's command and control monitor in the event of the commander's absence or incapacitation (ADMIN)

duty roster A listing of a unit's extra jobs and the personnel assigned to them on a rotating basis (ADMIN)

duty status Status of a soldier adjudged fit and able to do his or her assigned job (ADMIN)

DX Direct exchange (LOG)

DZ Drop zone (ABN)

DZSO Drop zone safety officer (ABN)

E

EA Engagement area (TAC)

EAC Echelons above corps (ADMIN)

eagle cocktail An expedient antiarmor device made of a sealed bag filled with gas and/or oil to which is attached an external, time-delayed charge, such as a grenade (ORD)

eagle fireball An expedient antiarmor device made of a can fitted with external grappling hooks and containing a gas/oil mixture and a white phosphorous grenade wrapped in det cord, actuated by a short time delay fuse (ORD)

E&E Evasion and escape (TAC)

early resupply The shipment of materiel between D-day and the beginning of planned resupply (LOG)

ECB Echelons corps and below (ADMIN)

ECC Equipment category code (LOG)

ECCM Electronic counter-countermeasures (COMMS)

echelon A level or subdivision of command; the physical siting or layering of successive or specialized command elements (ADMIN)

echelons above corps Command and control elements tailored to perform strategic-level missions (ADMIN)

ECM Electronic countermeasures (COMMS)

economic order quantity The computed optimum number of an item to order, based on its usage, cost, and delivery considerations (LOG)

economy of force The application of minimum essential combat power to secondary objectives (MS)

ECS Equipment concentration site (LOG)

ED Symbol for the blister agent ethyldichloroarsine (NBC)

E-date Effective date (ADMIN)

E-day The date on which a NATO exercise commences; a planning point within the wartime manpower planning system (ADMIN)

EDD Estimated delivery date (LOG)

edit To check a document for correctness (COLL)

EDRE Emergency deployment readiness exercise (TNG)

EECT End of evening civil twilight (ADMIN)

EEEEEE Prosign meaning error (COMMS)

EEFI Essential elements of friendly information (INTEL)

EEI Essential elements of information (INTEL)

EENT **1.** End of evening nautical twilight (ADMIN); **2.** Eyes, ears, nose, and throat (MED)

EER Enlisted evaluation report (ADMIN)

effective beaten zone The area on the ground where a high percentage of rounds fall (WPN)

effective damage The degree of destruction necessary to render a target inoperative (TAC)

effective range Maximum distance at which a weapon can reliably engage and destroy a target (WPN)

EFTO Encrypted for transmission only (COMMS)

EIB Expert Infantryman Badge (ADMIN)

electromagnetic camouflage The use of shielding, absorption, and/or enhancement techniques to obscure the electronic emanations of friendly forces (EW)

electromagnetic profile The pattern of electronic emissions produced by the equipment of a unit (EW)

electromagnetic pulse The disruption of the broadcast spectrum as a result of a nuclear blast (COMMS)

electromagnetic radiation Waves of energy that travel at the speed of light and can carry various types of information, e.g., radio waves, visible light, and infrared radiation (COMMS)

electromagnetic spectrum The complete range of the various types of electromagnetic radiation; *See* Table A-19 (COMMS)

electronic counter-countermeasures (ECCM) Those actions taken to ensure continued friendly ability to utilize the electromagnetic spectrum, despite enemy electronic countermeasures (EW)

electronic countermeasures (ECM) Those actions taken to deny or degrade the enemy's ability to effectively use the electromagnetic spectrum (EW)

electronic imitative deception The radiation of signals intended to resemble those of the enemy and thereby create confusion (EW)

electronic manipulative deception (EMD) The alteration of friendly signals and emissions to minimize revealing characteristics and introduce deceptive ones (EW)

electronic modulation The alteration of an electromagnetic wave to convey information (COMMS)

electronic simulative deception The radiation of signals intended to resemble those normally emitted by a given type of unit when, in fact, no such unit is present (EW)

electronics intelligence (ELINT) Evaluated information derived from noncommunications electromagnetic radiations (INTEL)

electronic warfare Military exploitation of the electromagnetic spectrum in support of conventional warfare (TAC)

electronic warfare support measures (ESM) Those actions designed to intercept, identify, and locate the source of enemy signals and emissions (EW)

elevation The angle between the bore of a weapon and the horizontal plane (ARTY)

elevation of security Minimum vertical angle at which a gun may be fired over friendly forces (ARTY)

elicitation The acquisition of information through conversation without disclosing that intent (INTEL)

ELINT Electronic intelligence (INTEL)

ELSEC Electronic security (COMMS)

EM Enlisted member (ADMIN)

embarkation The loading of troops and equipment onto ships or planes (LOG)

EMCON Emission control (COMMS)

emergency leave Absence authorized for soldiers with verified personal emergencies (ADMIN)

emergency priority The highest precedence of mission request; above urgent priority—reserved for situations such as response to an enemy breakthrough (ADMIN)

emergency risk A degree of anticipated danger from the effects of nuclear weapons, including casualties, temporary shock, and significant reduction of a unit's combat efficiency (NBC)

emission control (EMCON) Suppression of unintentional or unnecessary electromagnetic radiation to enhance operations and operational security (EW)

EMP Electromagnetic pulse (NBC)

emplacement A prepared position for a weapon (TAC)

encipher To convert plain text into an unintelligible form by use of a cipher system (COMMS)

enciphered code Plain text that is first encrypted by substituting arbitrary or symbolic words for the original text and then enciphered through character-for-character substitution (COMMS)

ENCOM Engineer command (ADMIN)

encrypt To convert plain text into unintelligible cipher or code text (COMMS)

end item An assembly of parts into a specialized machine or system, e.g., a tank, helicopter, or computer (LOG)

end of set marker The symbol "//"; used similarly to a period at the end of a sentence (COMMS)

end spell Indicator in a radio transmission of a written message that a spelling has been completed and that the remaining text represents words and phrases (COMMS)

endurance distance The range of movement of a vehicle without refueling (LOG)

enfilade A form of fire delivery in which the long axis of the beaten zone coincides with the long angle of the target (WPN)

engage **1.** To close with and battle the enemy (TAC); **2.** A fire control order that releases batteries to fire (AD)

engage hold A fire control order directing a battery to continue to track a target but hold fire; missiles in flight may be permitted to continue (AD)

engineering circuit An auxiliary channel dedicated to the maintenance and control of communications capabilities (COMMS)

engineers The combat branch and units responsible for construction and demolition in support of mobility, countermobility, and survivability (ADMIN)

enlisted person A soldier in the grades E-1 through E-9 whose term of service is governed by a contract (PERS)

ENTAC Entrance National Agency Check (INTEL)

entrenching tool A lightweight, collapsible shovel (INF)

entry group A cluster of closely related occupational specialties for which new soldiers are trained (TNG)

envelopment An offensive thrust beyond the enemy front toward their flanks and objectives to their rear, with the effect of surrounding them (TAC)

EO Executive order (LAW)

EOB Electronic Order of Battle (INTEL)

EOC Emergency operations center (ADMIN)

EOD Explosive ordnance disposal (ORD)

EOH Equipment on hand (LOG)

EOS Effect on system (LOG)

EOQ Economic order quantity (LOG)

epaulet Button-down shoulder strap; primarily ornamental (PERS)

EPW Enemy prisoner of war (LAW)

equilibrator A counterweight that facilitates the elevation of a gun tube (ARTY)

equipment modification list An amendment to a standard table of organization and equipment that tailors the table to the mission requirements of a given unit (LOG)

equipment serviceability criteria Checks done on important items of equipment to ensure that they are mission capable for at least 90 days with normal maintenance (LOG)

ERC Equipment readiness code (LOG)

ESC Equipment serviceability criteria (LOG)

escape chit A card offering a reward to anyone assisting the bearer of the card to safety (CA)

ESI Extremely sensitive information (INTEL)

ESM Electronic support measures (EW)

espionage The acquisition of information through covert or clandestine operations (INTEL)

essential elements of friendly information (EEFI) The key factors that an enemy commander is likely to want to know about friendly intentions and operations (INTEL)

essential items list Those items of materiel that are to be automatically returned to the national inventory control point if determined to be excess to the unit (LOG)

ETA Estimated time of arrival (ADMIN)

ETS Expiration, term of service (ADMIN)

evacuation hospital A semimobile medical treatment facility (MED)

evader Any individual isolated and eluding capture in unfriendly territory (LAW)

evaluation The appraisal of information being processed into intelligence as to its reliability and accuracy (INTEL)

evasion and escape Procedures and routes to be followed by personnel seeking to return to friendly lines from enemy-held territory (TAC)

evasion and escape net Organized assistance within enemy-held territory for personnel seeking to return to friendly lines (ADMIN)

EW **1.** Electronic warfare (TAC); **2.** Early warning (ADMIN)

excess property Property on hand that is not authorized by a higher headquarters (LOG)

exclusion area An area to which access is denied when access itself would compromise a security interest (ADMIN)

exclusive jurisdiction Agreement among state, national, and/or military authorities as to which is the sole civil and criminal authority for locations within the apparent jurisdiction of more than one of them (LAW)

execution planning The process of translating an operation plan or National Command Authority-ordered course of action into an operation order (ADMIN)

executive officer The second-in-charge of a command; the officer routinely charged with the execution of the commander's decisions (ADMIN)

executive order A directive with the force of law, issued by the President (LAW)

exempted station A military installation that reports directly to the Department of the Army General Staff rather than to a local area command (ADMIN)

exercise A simulated military operation, the purpose of which is to train personnel and test the feasibility of doctrine (TNG)

exercise term A two-word designation for training maneuvers; assigned to avoid confusion with actual operations (TNG)

exfiltration The surreptitious movement of soldiers out of an enemy-controlled area (TAC)

existence load Those items carried by a soldier, other than the fighting load and personal items normally carried, to enhance his or her survival capacity and protection (LOG)

expeditionary force A military unit organized and sent abroad to accomplish a specific objective (MS)

expendable property Materiel that is consumed through use and may then be dropped from accountability (LOG)

expert Highest qualification for accurate delivery of fire with an aimed weapon; above sharpshooter (PERS)

exploder A device that generates current in a firing circuit (ORD)

exploitation Offensive operation following a successful attack and designed to further disorganize the enemy in depth (TAC)

explosive charge One of two types of weapon blast—a propelling charge, which throws a projectile, or a bursting charge, which breaks the projectile apart for fragmentary, demolition, or chemical effect (ORD)

explosive ordnance Any bomb, warhead, charge, cartridge, round, or other munition (ORD)

explosive ordnance disposal The location, rendering safe, and removal of unexploded shells (ORD)

explosive train An arrangement of initiating and igniting elements designed to detonate a charge (ORD)

expulsion An order to citizens to vacate an area prior to hostilities (LAW)

extended defense a position defense characterized by a wide front, limited mutual support, depth of position, and withholding of a large reserve (TAC)

extent of damage Systematic reporting of destruction to a targeted area, expressed in percentages of 1,000-square-foot areas (INTEL)

extraction parachute A parachute designed to inflate and then pull cargo from an aircraft (LOG)

extraction zone An area for low-altitude air delivery of supplies/equipment (TAC)

eyes left (right) Two-part command directing soldiers in formation to turn their heads to look at the colors—a form of salute; cancelled by "ready, front" (DC)

EZ Engagement zone (AD)

FA Field artillery (ARTY)

FAASV The M992, a field artillery ammunition support vehicle (LOG)

FAC Forward air controller (ARTY)

facing distance The measured front required for an individual soldier in formation (DC)

facsimile The telecommunication of visual images; commonly called remote copying or fax (COMMS)

FAD Force activity designator (LOG)

FADAC Field artillery digital automatic computer (ARTY)

Fade Out Exercise term for DEFCON 5, a state of alert (STRAT)

fail safe A device that disarms a weapon or munition in the event of a malfunction that would make it prone to accidental launch or detonation (ORD)

fair wear and tear Basis for issue of new materiel when the old item is worn out and turned in (LOG)

fall in A combined command that directs troops to assemble a formation (DC)

fall out A combined command that directs troops to disassemble a formation (DC)

fallout Radioactive debris that rains to the earth's surface following a nuclear burst (NBC)

fallout contours Lines on a map that connect points of equal radiation intensity (NBC)

fallout prediction An estimate of the dispersion of militarily significant radioactive material resulting from a nuclear burst; variables include type and size of burst and weather conditions (NBC)

fallout safe height of burst The altitude at or above which no militarily significant fallout will be produced by a nuclear airburst (NBC)

FALOP Forward area limited observing program (INTEL)

false flag Operation that presents itself as aligned or sympathetic to forces opposing its true allies (UW)

FAO **1.** Finance and accounting office (ADMIN); **2.** Foreign area officer (PERS)

FARP Forward area refueling point (LOG)

FASCAM Family of scatterable mines (ORD)

Fast Pace Exercise term for DEFCON 2, a state of alert (STRAT)

fax Facsimile (COMMS)

FC **1.** Field circular (TNG); **2.** Finance Corps (ADMIN)

FCL Fire coordination line (TAC)

FCX Fire coordination exercise (TNG)

FDC Fire Direction Center (ARTY)

FEBA Forward edge of the battle area (TAC)

Federal Stock Number Obsolete (as of 30 September 1974) identification number for an item in the Federal Supply Classification system (COLL)

federal supply class Broad categorizations used in the standardization, screening, and management of materiel; *See* Table A-10 (LOG)

feint A show of offensive force, such as a limited attack, intended to distract enemy attention from a main attack elsewhere or to test enemy response (TAC)

ferret Any platform equipped to detect and/or analyze electromagnetic radiation (EW)

FES Field entry standard (TNG)

FEZ Fighter engagement zone (AD)

FFZ Free fire zone (TAC)

FID Foreign internal defense (UW)

field army Administrative and/or tactical unit consisting of a headquarters with organic service and/or support troops, at least two corps, and a variable number of divisions (ADMIN)

field artillery The combat branch and units responsible for the employment of cannons, rockets, or surface-to-surface missiles and their supporting materiel and personnel (ADMIN)

field artillery survey The topographic intelligence on the exact location and elevation of unobserved targets sufficient for targeting (ARTY)

field circular Publication containing updated doctrine prior to its inclusion in field manuals (TNG)

field exercise The practice deployment of friendly troops in a simulated mission against a notional enemy (TNG)

field expedient An improvised remedy or solution (COLL)

field flag A nylon/wool national flag measuring 6 feet, 8 inches, by 12 feet that is displayed with the positional field flag (DC)

field fortifications Shelter or defensive emplacements of a limited, temporary nature, capable of erection with minimal engineer supervision (TAC)

field grade officer An officer holding the rank of major, lieutenant colonel, or colonel (ADMIN)

field manual A basic reference document containing official doctrine tactics and operational information (TNG)

field marker Message text symbol ("/") that indicates the beginning of a data field (COMMS)

field of fire Sector that can be effectively covered by fire from weapons in a given position (TAC)

field operating activity Unit with a mission to execute policy; the unit would be required even in the absence of the headquarters it reports to (ADMIN)

field service regulations The 100-series of manuals that present the fundamental doctrine and policy to be used in conducting operations at division level and higher (TAC)

field station A fixed strategic site for the interception of radio communications (INTEL)

field strip To partially disassemble a weapon for the purpose of cleaning and minor repair (WPN)

field train Combat service support elements such as administrative, heavy maintenance, and food and water supply elements; units not directly supporting combat operations (LOG)

fighting load All supplies and equipment to be carried by an infantryman in accomplishment of an operational mission (LOG)

FIIN Federal item identification number (LOG)

file A column of soldiers (or other elements) in which all face the same direction, one behind the other (DC)

film badge A device for recording a soldier's exposure to radiation (NBC)

final coordination line A recognizable line at which forwardly deployed maneuver elements request their last adjustment of supporting fires (TAC)

final protective fire A barrier of on-call fire designed to stop enemy crossing of defensive areas (TAC)

final protective line Defensive front at which the enemy is to be stopped by interlocking fire from all available weapons (TAC)

Finance Corps The combat service support branch and personnel responsible for budget and payroll administration (ADMIN)

fine sight Adjustment of a gun's aiming device so that only the tip of the front sight can be seen through the rear sight when the gun is properly aimed at the target (WPN)

fire and forget Weapons that are smart but self-guided, not requiring continued target input (WPN)

fire and maneuver Method of attack in which one element moves under the cover of suppressive fire from another element (TAC)

fire control equipment Computers, range finders, sighting devices, and communications and adjustment mechanisms used to direct guns and missiles (ARTY)

fire coordination area A sector into which fire cannot be delivered without clearance from higher command levels (TAC)

fire direction Tactical command and control of targeting, allocation of ammunition, and distribution of fire (ARTY)

fire for effect Fire delivered for a destructive purpose rather than for ranging or testing enemy reaction (ARTY)

fire mission Tactical artillery support as part of a larger operation (ARTY)

fire plan The coordination of a unit's weapons to ensure coverage of target areas without unnecessary duplication (TAC)

fire support Artillery and close air support in assistance to infantry and armor units (TAC)

fire support coordination The planning and execution of artillery and close air support assets to maximize their efficiency (ARTY)/(TAC)

fire support coordination line (FSCL) A line along prominent terrain features dividing forward sectors into which (a) fire can be delivered on enemy targets only with the ground force commander's approval or (b) fire support can be delivered at the discretion of the support elements (TAC)

fire trench A ditch from which soldiers can deliver small arms force while presenting a limited target to the enemy (INF)

fire unit A unit whose fire is controlled in battle by a single commander (TAC)

firing chart A map or chart that depicts the relative positions of batteries, base points and lines, checkpoints, targets, and other data critical to fire control (ARTY)

firing circuit An electrical network connecting prepositioned charges (WPN)

firing data Targeting information necessary to fire a gun on a particular objective (ARTY)

firing device Trigger for a booby trap (WPN)

firing jack Device that stabilizes and adjusts certain mobile artillery pieces while in the firing position (ARTY)

firing pin The part of the firing mechanism that strikes the primer, igniting the propellant (WPN)

firing table Compilation of the data required to accurately employ a gun under standard conditions and the corrections necessary for nonstandard conditions (ARTY)

first defense gun Machine gun positioned to cover enemy avenues of approach from initiation of attack to the front lines of the defensive positions (TAC)

first destination The point in the supply system where materiel is initially received in the form required for use (LOG)

first lieutenant (1LT) A commissioned officer; pay grade O-2 (PERS)

first light The earliest moments of dawn (ADMIN)

first sergeant (1SG) Noncommissioned officer; senior enlisted soldier in a company-sized unit; pay grade E-8 (PERS)

fiscal station A numbered installation authorized to perform certain appropriation, fund accounting, and materiel furnishing functions (ADMIN)

fiscal year The basic budgetary planning year, beginning 1 October and ending 30 September (ADMIN)

fishbowl An interrogation room that permits observation from the outside via mirrored glass (INTEL)

FISINT Foreign Instrumentation Signals Intelligence (INTEL)

FIST Fire support team (ARTY)

five quarter A 1.25-ton truck (LOG)

fix **1.** A geographical position (NAV); **2.** The intersection of three or more lines of bearing (EW)

fixed ammunition Ammunition in which the cartridge case comes attached to the projectile prior to firing (ORD)

fixed call sign/frequency Contact data that remain unchanged for at least one year (COMMS)

fixed fire Fire that is neither searching nor traversing, but is delivered on a single point (TAC)

fixed-wing Nonhelicopter aircraft; airplanes (AVN)

flag officer A general officer; one who is authorized to fly his or her own starred flag (PERS)

flail tank A minesweeping armored vehicle that utilizes chains on a roller to detonate mines (ARM)

flank **1.** The side of an element or formation (TAC); **2.** A preparatory command preceded by "right" or "left" directing all elements of a moving formation to execute a 90-degree turn to the specified direction on the command of execution, march (DC)

flanking attack An offensive maneuver directed at the side, rather than the front, of an enemy element (TAC)

flash A colored unit-designating patch worn on a beret (PERS)

flash blindness Temporary or permanent impairment of vision resulting from intense bursts of light (NBC)

flash interrupt message (ZI) Handling precedence above flash and below flash override (COMMS)

flash message (ZZ) Handling precedence above immediate and below flash interrupt; reserved for messages regarding initial enemy contact or extremely urgent and brief operational messages (COMMS)

flash override message (WW) Highest handling precedence above flash interrupt (COMMS)

flash suppressor Device at the muzzle of a gun's barrel designed to minimize the burst of visible light that occurs when firing (WPN)

flash-to-bang time The interval between the sight and the sound of the detonation or firing of a weapon or projectile (TAC)

flash tube A path for transmission of a spark or flame from a detonating device to an explosive charge (WPN)

fléchette A small, antipersonnel, fin-stabilized missile (ORD)

Fleer 9 The AN/FLR-9(V) fixed, strategic antenna array used for interception and direction-finding (EW)

fleeting target A moving target that remains within range for too short a time to permit deliberate fire adjustment (ARTY)

flexible response The ability of a military force to adapt and respond to the level of enemy threat in an appropriate manner (MS)

flex-x Explosive demolition charge in flexible sheets (ORD)

flipper A device for ejecting scatterable mines (WPN)

FLOT Forward line of own troops (TAC)

FM **1.** Frequency modulation (COMMS); **2.** Prosign meaning from (COMMS); **3.** Field manual (TNG)

FMC Fully mission capable (LOG)

FO Forward observer (TAC)

fog of war Confusion; the lack of certainty during operations due to degraded communications and the fluidity of a situation (MS)

fog oil Petroleum specially formulated for maximum smoke generation (LOG)

FOIA Freedom of Information Act (LAW)

follow-on Successor (COLL)

FOMCAT Foreign materiel catalog (LOG)

force accounting system The comprehensive data processing system used to manage and control the forces of the U.S. Army (ADMIN)

force augmentation The mobilization of high priority Reserve units and individual soldiers to bring active duty units to full strength for a specified mission objective (ADMIN)

forced crossing The fording or swimming of a water obstacle while under enemy fire (TAC)

forced issue Involuntary distribution of materiel (LOG)

force tabs Deployment schedules for major combat units within a theater (STRAT)

fording Crossing a body of water by wading personnel or vehicles (TAC)

foreign instrumentation signals intelligence Intelligence derived from intercepted aerospace telemetry (INTEL)

foreign internal defense U.S. assistance to civilian and military agencies of another government to thwart subversion, lawlessness, and insurgency (UW)

fork The change in a gun's elevation that shifts the center of the rounds' impact by four probable errors (ARTY)

formal accountability The requirement to maintain a stock record or commissary account of all property that is not specifically exempted from accountability (LOG)

formation The orderly arrangement of a unit's elements (DC)

FORSCOM United States Army Forces Command (ADMIN)

fort A permanent, year-round military installation (ADMIN)

forward Toward or close to the enemy (TAC)

forward air control The coordination of aviation assets in close support of ground forces (AVN)

forward area An area near enemy contact (TAC)

forward echelon The sector nearest the enemy, which contains those forces concerned with tactical operations (TAC)

forward edge of the battle area (FEBA) That sector closest to enemy positions where friendly combat and maneuver echelons are deployed (TAC)

forward line of own troops (FLOT) The friendly combat echelon closest to enemy positions at a given time (TAC)

forward observer A soldier within visual range of artillery round impact who reports the result and placement of that fire back to the gun or fire-control element (ARTY)

forward slope Terrain that descends toward the enemy (TAC)

forward tell The communication of information to a higher command level (INTEL)

fougasse A mine designed to blow debris or shrapnel in a specific direction (ORD)

found on post The supply designation that accounts for materiel, the accountable holder of which is unknown (LOG)

FOUO For official use only (ADMIN)

fourragère A braided shoulder cord representing a foreign decoration (PERS)

foxhole Any small pit providing partial cover as a firing position (TAC)

FPF Final protective fire (TAC)

FPO Fleet Post Office (ADMIN)

FR Prosign meaning for (COMMS)

fragmentary order An order format for issuing supplemental instructions on a current operation; "frag" is also used (ADMIN)

FRAGO Fragmentary order (ADMIN)

fraternization Socialization between ranks or sexes to the detriment of either a soldier's duties or the unit mission (LAW)

free drop Air delivery of supplies without use of parachutes (ABN)

free-fire area A sector into which fire can be delivered without coordination (TAC)

free gun A combination of searching and traversing fire in which the rounds are distributed in both width and depth (TAC)

free issue Supply of materiel without accountability or financial charge to the using unit (LOG)

free maneuver Practice exercise in which the opposing forces' actions are determined primarily by their own initiative (TNG)

free net An uncontrolled communication circuit in which any station can contact any other without having to obtain permission (COMMS)

free rocket A missile not subject to control after launch (WPN)

freq Frequency (COMMS)

frequency hopping Radio equipment that automatically varies the linking frequency for security purposes (COMMS)

frequency modulation The variation of the wavelength of a broadcast radio signal to enable it to carry information (COMMS)

friendly Nonenemy; all allied and joint U.S. forces (COLL)

FROG Free rocket over ground (WPN)

front **1.** Toward the enemy (TAC); **2.** The line of lateral contact between opposing forces (TAC)

frontage The lateral distance between the forwardmost flanks of a tactical element (TAC)

frontal attack An offensive action directed at the center of the enemy's forward lines (TAC)

front and center Combined command directing a soldier in formation to report to the formation's command element (DC)

front-leaning rest The initial pushup position—the "up" position (TNG)

frostbite Freezing damage to exposed skin; symptoms include discoloration, pain, numbness, and frozen solid tissue; to treat, shelter from continued freezing and very gradually warm the tissue (MED)

FS Fire support (ARTY)

FSAO Family Service and Assistance Officer (ADMIN)

FSCL Fire support coordination line (TAC)

FSCOORD Fire support coordinator (ARTY)

FSE Fire support element (ARTY)

FSN Federal stock number (LOG)

FSP Faculty security profile (ADMIN)

FSU Field storage unit (LOG)

FTUS Full-time unit support (ADMIN)

FTX Field training exercise (TNG)

full bird A full colonel; pay grade O-6 (COLL)

full mobilization Activation of all Reserve units and available individuals to meet the national security requirements resulting from an external threat (STRAT)

functional area A grouping of officers by advanced specialized skills other than their service or branch (TNG)

functional files system (TAFFS) A standardized sequence of and instructions for the classification, filing, reference, and disposition of a unit's documents (ADMIN)

functional training Training that fulfills a required skill that is not part of the military occupational specialty system (TNG)

fund citation An alphanumeric code that designates the monetary account to which an authorized expense is to be charged (ADMIN)

fund code A two-position designator used to match requisitions with their payment source within the Military Standard Requisitioning and Issue Procedures system (LOG)

funds responsibility The control of authorized funds and preparations of budget estimates (ADMIN)

fuse A device in the form of a flammable or explosive cord used to detonate, remotely or with a time delay, an explosive charge (ORD)

fuze A device with explosive components that initiates detonation of a larger bomb or munition (ORD)

FWT Fair wear and tear (LOG)

FY Fiscal year (ADMIN)

FYI For your information (COLL)

G

G-1 Assistant Chief of Staff for Personnel (ADMIN)

G-2 Assistant Chief of Staff for Intelligence (ADMIN)

G-3 Assistant Chief of Staff for Operations/Plans (ADMIN)

G-4 Assistant Chief of Staff for Logistics (ADMIN)

G-5 Assistant Chief of Staff for Civil Affairs (ADMIN)

GA Symbol for the nerve agent Tabun (NBC)

Galaxy The fixed-wing C-5A, the largest transport aircraft in the U.S. inventory (AVN)

gallery Tunnel (ENG)

gamma radiation Nuclear emissions of short range capable of producing internal injury (NBC)

garble An error in a message text that renders that portion of the message not understandable (COMMS)

garrison The administrative stationing of troops in a secured or fortified area (LOG)

garrison cap Flat, foldable dress hat (PERS)

garrison flag A nylon/wool national flag measuring 20 by 38 feet that is displayed on holidays and important occasions (DC)

gator An air-deliverable, scatterable mine system (WPN)

GB Symbol for the nerve gas sarin (NBC)

GD Symbol for the nerve agent soman (NBC)

GEMSS Ground-emplaced mine scattering system (ORD)

GEN General (PERS)

general A commissioned officer; pay grade O-10 (PERS)

general court-martial A court of at least five members, plus a military judge, authorized to try any offense under the Uniform Code of Military Justice (LAW)

general discharge Characterization of a soldier's term of service as honorable, with service at least satisfactory but not otherwise qualifying for an honorable discharge (ADMIN)

general hospital A fixed medical facility, organized under a table of organization and equipment, providing specialized and definitive treatment (MED)

general orders Standard guidance for all interior guards in the performance of their duty as follows:

1. I will guard everything within the limits of my post and quit my post duty when properly relieved

2. I will obey my special orders and perform all my duties in a military manner

3. I will report violations of my special orders, emergencies, and anything not covered in my instructions to the commander of the relief (TAC)

general staff Those officers who function in support of a general command level (division); designated by G-series sections (ADMIN)

general support Assistance to a force as a whole rather than to a particular subdivision of the force (ADMIN)

general war Conflict between major powers whose national survival is at stake, employing all resources at their disposal (MS)

generation The successive reproduction of imagery, which entails increasing loss of detail (INTEL)

Geneva Conventions A series of agreements among nations that define the rules and limitations under which armed conflict will be conducted, particularly regarding limits of force and targets and treatment of prisoners of war and noncombatants (MS)

geographic coordinates Lines of latitude and longitude that define a position on the earth's surface (NAV)

georef A position reference system capable of worldwide expression of latitude and longitude on any type of map projection; coordinates consist of four letters and four numbers (NAV)

GFT Graphical firing tables (ARTY)

ghost A soldier who is frequently absent (COLL)

GHz Gigahertz (EW)

GI **1.** Government issue (COLL); **2.** A soldier (COLL); **3.** To clean (COLL)

gig An infraction or shortcoming; to find fault (COLL)

gig line The vertical line that should be formed by the shirt opening, belt buckle, and trouser fly (COLL)

gisting The copying of a short version of the important points of an intercepted message as it is being broadcast (INTEL)

glacis plate Sloping armor around a tank's turrets and hatches (ARM)

GM angle Grid magnetic angle (NAV)

GMT Greenwich mean time (ADMIN)

GN Grid north (NAV)

go **1.** To pass; a passing grade (TNG); **2.** Equipment status—ready (LOG); **3.** An elevation of open terrain mobility (TAC)

GOCOM General officer command (ADMIN)

Good Conduct Medal Decoration recognizing three years of continuous active service conducted in a personally exemplary manner; precedence—follows U.S. nonmilitary decorations and precedes service medals (PERS)

GP General purpose (LOG)

GPM Groups per minute (COMMS)

GPMG General purpose machine gun (WPN)

GPW Geneva Convention Relative to the Treatment of Prisoners of War, 12 August 1949

GPW 1929 Geneva Convention Relative to the Treatment of Prisoners of War, 27 July 1929

GR Prosign meaning group count (COMMS)

gradability Ability of a vehicle to handle the slope of terrain; measured in percentages (LOG)

grade Level of base pay corresponding to a soldier's rank (ADMIN)

granulation The size and shape of the grains of a propellant charge (ORD)

grape shot Field-expedient shrapnel, such as nuts and bolts (WPN)

graphic training aids Charts, posters, and training devices used to instruct soldiers either individually or in the classroom (TNG)

graves registration The process of identification, removal, and burial of the dead from the battlefield and the handling of their effects (LOG)

gray propaganda Propaganda, the source of which is unclear (UW)

grazing fire Fire parallel to the ground at an elevation of no more than one meter (TAC)

grazing herd message An alert notification that is only a test of the alert system (ADMIN)

grease gun The M3A1 .45 caliber submachine gun (WPN)

greenspeak The use of Army slang or jargon or excessive verbosity to the detriment of clear communication (COLL)

Greenwich mean time (GMT) The basis for world standard time; the current time at Greenwich, England, which serves as the world standard time basis; also known as zulu time (ADMIN)

grenade A small bomb designed to be thrown or launched by an infantryman in close contact with the enemy (ORD)

grid azimuth A horizontal angle, measured clockwise from grid north (NAV)

grid convergence The horizontal angular difference between true north and grid north (NAV)

grid coordinates Alphanumeric designation of a set of perpendicular lines that identifies a point on a map (NAV)

grid line A north/south or east/west coordinate of a map (NAV)

grid magnetic angle The difference, in degrees, between grid north and magnetic north, measured clockwise from grid north (NAV)

grid north The direction indicated as north by a map coordinate system (NAV)

gross requirements Total materiel needs, including initial issues, maintenance float, operational projects, pipeline quantities, and post D-day consumption (LOG)

gross weight Total weight; weight of a vehicle, including everything on it—load, fuel, passengers, and equipment (LOG)

ground guide Dismounted soldier who assists in the safe maneuver of vehicles (LOG)

grounding The establishment of an electrical path between equipment or vehicles and the earth to prevent the buildup of a potentially damaging electrical charge (MAINT)

ground observer team A small detachment deployed to observe and report on enemy aircraft movements (AD)

ground wave Radio signal that tends to hug the ground, but only to the horizon; it has virtually no sky wave component (COMMS)

ground zero The location on the earth's surface of the center of a nuclear detonation, or the point directly below an airburst (NBC)

group (GP) **1.** A unit of variable size but normally designating two or more battalions (ADMIN); **2.** Two or more links operating under common control in a network (COMMS)

grreg graves registration (LOG)

GRU The Soviet military intelligence agency—similar to the Defense Intelligence Agency (DIA) (INTEL)

GS General support (ADMIN)

G Series A family of highly toxic chemical nerve agents (NBC)

GSR **1.** Ground surveillance radar (INTEL); **2.** General support, reinforcing (ADMIN)

GSU General support unit (ADMIN)

GT General technical; an approximate measure of an individual's mental ability (ADMIN)

GTA Graphic training aid (TNG)

GTR Government travel request (LOG)

guard A supporting element that protects the main force through observation and fire (TAC)

guarded frequency A portion of the radio communications spectrum that is not to be jammed because friendly forces are monitoring enemy use of it (EW)

Guardrail The AN/USD-9 fixed-wing aircraft capable of interception, direction-finding, control, and reporting functions (EW)

guerrilla forces Combatants not employed by an organized national army—irregular forces (PERS)

guerrilla warfare Military operations conducted in hostile territory by irregular forces (MS)

guidance **1.** Policy or instruction with the force of an order (ADMIN); **2.** The direction of a projectile to its target (WPN)

guided missile A self-propelled projectile, the trajectory of which is controlled during flight (WPN)

guide left (right) Combined command directing elements of a moving formation to maintain their side interval in relation to the element to their immediate left or right; the leftmost or rightmost element follows the required terrain contour (DC)

guidon Dovetail-shaped flag authorized to identify detachments and separate platoons of 30 or more soldiers (ADMIN)

gun A long-barreled cannon with a relatively low angle of fire and a high muzzle velocity (ARTY)

gun carriage The supporting mechanism for a large cannon, which may include the elevating and traversing devices (ARTY)

gun density The number of guns that can fire on a given target (TAC)

gun displacement Distance between the base piece of a battery and any other gun (ARTY)

HAHO High altitude/high opening (ABN)

half left (right) Preparatory portion of a two-part command directing troops in a moving formation to execute a 45-degree turn on the command of execution, march (DC)

half step **1.** To march with a 15-inch step (DC); **2.** To do anything too slowly (COLL)

half track Vehicle, often armored, propelled by rear, complete band tracks and steered by front wheels (LOG)

HALO High altitude/low opening (ABN)

handover The transfer of responsibility for maintaining combat initiative from one unit to another (TAC)

handover line Battle management control feature, preferably defined by terrain features, along which control of combat operations is passed from one unit to another (TAC)

hand receipt A document used to record personal accountability for supplies or equipment (LOG)

hang fire The malfunction of a firing system (ARTY)

harassment An operation designed to disturb or disrupt the enemy rather than to inflict serious damage (TAC)

hardball A paved road (COLL)

hardened site A defensive position fortified to protect against direct artillery hits or the blast of a nuclear detonation (TAC)

hardship tour An overseas duty assignment where civilian community facilities are minimal and family members are not permitted to accompany the soldier (ADMIN)

hard stripe Sergeant or corporal E-4–E-9, as opposed to a specialist (COLL)

hash marks Service stripes (COLL)

hasty An operation conducted with little preparation under battle conditions (TAC)

hasty attack An offensive operation in which preparation time is traded for an immediate opportunity to exploit temporary enemy vulnerability (TAC)

hasty defense A minimal defense organized during contact with the enemy or while contact is imminent (TAC)

Hawk A mobile, medium-range air defense artillery missile system that provides very-low-altitude to medium-altitude coverage in conjunction with ground forces (AD)

HD Symbol for the blister agent distilled mustard (NBC)

HE High explosive (ORD)

headcount The number of troops fed at a meal or the soldier doing the counting (LOG)

heading Direction of travel, normally expressed in degrees clockwise from true north (NAV)

headquarters The administrative elements of a command (ADMIN)

Headquarters, Department of the Army The highest-level headquarters within the Army, consisting of the Army Secretariat, the Army General and Special Staffs, and designated staff support agencies (ADMIN)

head shed Headquarters (COLL)

head space Distance between the face of a fully closed bolt and the cartridge seating shoulder of the chamber (WPN)

heads up A warning order; any advice of future operations (COLL)

head-up display The projection of flight control data upon the pilot's forward field of view (AVN)

HEAT High explosive, antitank (ORD)

heat cramps Muscle cramps and sweating caused by electrolyte imbalance; treatment—move to shade, drink water, loosen clothing (MED)

heat exhaustion Extreme loss of water through perspiration that results in headache, weakness, dizziness, nausea, chills, heat cramps, and/or rapid breathing; treatment—move to shade, drink water, loosen clothing, elevate legs, place on light duty status, and monitor symptoms (MED)

heat stroke Failure of the body's cooling system due to extreme heat exposure, resulting in flushed, hot, dry skin, weakness, confusion, heat cramps, seizures, rapid and weak respiration and/or pulse, and/or unconsciousness; treatment—cool by all means possible, massage extremities, drink water, elevate legs, monitor for development of life-threatening symptoms, and seek medical aid (MED)

heat tabs Any solid or pelletized chemical used to start a fire or cook in damp or windy conditions, or when wood is not available to burn (INF)

heavy Descriptive of a unit with a high percentage of armored vehicles (COLL)

heavy artillery Cannons and ammunition of a caliber between 161 mm and 210 mm (ARTY)

heavy drop Delivery of large items of materiel by fixed-wing aircraft through use of extraction parachute and/or conveyor system (LOG)

heavy level of operations Combat employing at least 60 percent of available maneuver echelons and all available fire support over an extended period, including possible employment of higher echelons to ensure mission success (TAC)

heavy machine guns Automatic guns .30 caliber and larger (WPN)

helicopter wave rendezvous An air control point for assembly of helicopters prior to execution of a mission (AVN)

Hellfire An air-launched, laser-guided, antiarmor missile (WPN)

Hell on Wheels The Second Armored Division (ARM)

HEMMS Hand-emplaced minefield marking set (ORD)

Hercules The C-130, a fixed-wing, medium-range transport aircraft (LOG)

HET Heavy equipment transporter (LOG)

HEXJAM Small, disposable communications jammers that can be distributed by infantry (EW)

HF High frequency (COMMS)

HHC Headquarters, headquarters company (ADMIN)

HHD Headquarters, headquarters detachment (ADMIN)

H-hour The time at which an operation is to commence; a planning point within the wartime manpower planning system (ADMIN)

high-angle fire Fire that decreases in range as the gun's elevation increases (ARTY)

higher Higher level of command (COLL)

high explosive Material that changes from a solid to a gaseous state with great destructive power at a rate of about 1,000 meters per second (ORD)

high grade cryptosystem Encipherment designed to resist solution indefinitely (COMMS)

high nuclear yield Nuclear weapon burst that generates 50–500 kilotons of blast energy (WPN)

high payoff target An enemy position, unit, or item of equipment, the destruction of which is considered critical to friendly operations (TAC)

high port To hold a rifle diagonally in front of the torso while running or jumping (TAC)

high speed The most advanced, technical, or up-to-date (COLL)

high value target An enemy position, unit, or item of equipment judged by friendly intelligence as critical to enemy operations (TAC)

high velocity **1.** Projectiles with muzzle velocities from 3,000 to 3,499 feet per second (ARTY); **2.** Tank cannon projectile velocities from 1,550 to 3,350 feet per second (ARM); **3.** Small arms round velocities from 3,500 to 5,000 feet per second (WPN)

high velocity drop An airdrop in which the falling speed is slowed by a parachute but still exceeds 30 feet per second (ABN)

HIMAD High- to medium-altitude air defense (AD)

hip pocket training Basic instructional material that is kept readily available for unanticipated training opportunities (TNG)

historical cost The actual expenditure for materiel, which, if not known, can be estimated by using the current cost and a deflator index (LOG)

HL Symbol for the blister agent mustard lewisite (NBC)

HM HM Prosign meaning emergency silence (COMMS)

HMMWV High mobility, multipurpose wheeled vehicle (LOG)

HN-1, -2, -3 Symbols for the nitrogen mustard family of blister agents (NBC)

HNS Host nation support (STRAT)

hold fire An emergency fire control order to cease fire and destroy missiles already in flight (AD)

holding attack Offensive action meant to occupy the attention of the enemy, freeze them in place, and deceive them as to the location of the main attack (TAC)

Hollywood jump A parachute jump made without personal equipment (ABN)

homing The direction of a mobile station or weapon to a specified point (WPN)

honorable discharge Characterization of a soldier's term of service as honest and faithful, with conduct ratings of at least good, efficiency ratings of at least fair, no convictions by a general court-martial, and no more than one conviction by a special court-martial (ADMIN)

HOR Home of record (ADMIN)

horizontal action mine A mine designed to produce its destructive force parallel to the ground (ORD)

horizontal error A weapon's average distance error in hitting a target, half of the rounds hitting closer and half farther from a target (WPN)

hospital center An administrative unit to which hospitals and other medical units are assigned for command and control purposes (MED)

hot An area under fire or where combat is taking place (TAC)

HOTX Hands-on training exercise (TNG)

howitzer A cannon capable of high-angle fire (ARTY)

HPT High-payoff target (ARTY)

HS Highly sensitive (INTEL)

Huey The UH-1 helicopter—a light, single-rotor transport and attack support aircraft; officially designated as the Iroquois (AVN)

hull The armored body of a tank (ARM)

hull down Defensive position in which the body of an armored vehicle is shielded by earthworks from direct fire (ARM)

HUMINT Human intelligence (INTEL)

HVAPDS Hypervelocity armor-piercing discarding sabot (ORD)

HVT High value target (ARTY)

HW Prosign meaning herewith (COMMS)

hydrogen cyanide (AC) A toxic blood agent; decontamination not necessary in an open space (NBC)

hydropneumatic Liquid-gas charged device designed to absorb the recoil of certain guns (ARTY)

hypervelocity **1.** Muzzle velocity in excess of 3,500 feet per second (ARTY); **2.** Small arms muzzle velocity in excess of 500 feet per second (WPN); **3.** Muzzle velocity of tank projectiles in excess of 3,350 feet per second (ARM)

hypothermia Life-threatening loss of body heat; symptoms include lowered body core temperature, shock, unconsciousness, and uncontrolled movement but cessation of shivering; treatment includes drying, warming, and seeking medical aid (MED)

Hz Hertz (COMMS)

I&W Indications and warnings (INTEL)

IAW In accordance with (ADMIN)

ICBM Intercontinental ballistic missile (WPN)

ICD Imitative communications deception (EW)

icecrete A frozen mixture of sand, gravel, and water (LOG)

icemining The breaking of surface ice on a river or lake to prevent enemy passage (TAC)

ICM Improved conventional munitions (ORD)

ICOM Improved conventional mines (ORD)

IDAD Internal defense and development (UW)

IDT Inactive duty training (TNG)

IED Initiative electronic deception (EW)

IEW Intelligence electronic warfare (TAC)

IEWSE Intelligence electronic warfare support element (EW)

IFF Identification, friend or foe (WPN)

IG Inspector general (ADMIN)

igniter A device used to initiate a firing circuit (WPN)

II Prosign meaning separator sign (COMMS)

ILO In lieu of (ADMIN)

image interpretation The analysis of (primarily overhead) imagery or photographs of enemy positions to derive intelligence from them (INTEL)

imagery The visual representation of objects through photographs, infrared detection, light of any wavelength, or optical enhancement (INTEL)

imagery pack An assembly of imagery from various sources covering a single target (INTEL)

IMC International Morse code (COMMS)

I method Type of message transmission in which the receiving station does not acknowledge receipt and is responsible for correct reception (COMMS)

IMI Prosign meaning question mark; interrogative (COMMS)

IMINT Imagery intelligence (INTEL)

imitative electronic deception (IED) The transmission of electronic emissions that mimic enemy communications and noncommunications emissions (EW)

immediate action **1.** The clearing of a stoppage in a weapon (WPN); **2.** To check a wounded person's vital signs and attempt to restore them, if necessary (MED)

immediate air support Air support in response to battlefield necessities that cannot be planned for (TAC)

immediate message (OO) Handling precedence above priority and below flash; indicates a situation of grave importance to national security (COMMS)

immediate reenlistment An enlistment within 24 hours after an expiration of term of service under a prior enlistment (ADMIN)

immediate smoke An unplanned mission to create a point of obscuring smoke of very short duration (ARTY)

immersion proof Equipment that can be submerged under three feet of water for two hours and then operate normally immediately after removal (LOG)

immersion syndrome Gradual destruction of body tissue due to prolonged submersion in cold water; symptoms include alternate pain and numbness, discoloration, bleeding, swelling, and gangrene; treatment involves drying, gradual warming, and elevation—do not massage or apply direct heat (MED)

impact area A bounded area designated as the landing area for projectiles from a training range (TNG)

imprest fund A relatively small cash fund for authorized purchases of supplies and nonpersonal services (ADMIN)

INA Information not available at the unclassified level (ADMIN)

inactive duty training Reserve component training conducted at regularly scheduled unit training assemblies, additional training assemblies, and/or equivalent training not on active duty (TNG)

inactive National Guard Guardsmen temporarily assigned to Ready Reserve status as a result of their inability to participate in training (ADMIN)

inactive status list Standby Reservists who are unable to participate in training but are retained for contingency duty (ADMIN)

incapacitating agent Chemical that temporarily renders personnel incapable of functioning or performing their duties (NBC)

incendiary Fire-inducing ammunition (ORD)

incentive pay Money paid in excess of basic pay for performance of hazardous duty or to maintain specialized skills (ADMIN)

indefinite call sign A radio call up that is directed to an unspecified station or group of stations (COMMS)

indeterminate change of station Assignment to a temporary duty station, pending permanent assignment to a duty station to be determined (ADMIN)

index contour line Every fifth contour line on a military map; printed in boldface and labeled with the indicated elevation (NAV)

index error A measurement error caused by the initial misalignment of an instrument (LOG)

indications Observations and information that tend to reveal an enemy's potential course of action (INTEL)

indicator regulator Instrument that computes and displays firing data (ARTY)

indigenous personnel Natives (COLL)

indirect fire Fire on a target that is obscured from sight at the weapon being used (WPN)

indirect illumination Battlefield lighting employing diffusion or reflection in order to mask the position of the light source (TAC)

indirect laying Aiming a gun by means other than sighting directly on the target (ARTY)

individual Ready Reserve A reservist who is not assigned to a Reserve unit and not on active duty (ADMIN)

indorsement Reply to or transmittal of military correspondence (ADMIN)

indorser The second-level evaluator (above the rater) in an efficiency assessment system (ADMIN)

induced radiation Transferred radioactivity that persists after exposure to a nuclear detonation (NBC)

induction field locator Radio transmitter inserted into airdropped bundles to permit homing on their location (ABN)

inert Nonoperational; a mock-up or dummy version for instructional use (COLL)

inertial guidance Adjustment of a missile's course based on its own internal sensing and balancing of acceleration and direction (WPN)

infantry The combat branch and units trained to move and fight primarily on foot (ADMIN)

infiltration The undetected movement of a small force or of individuals through enemy-held territory (TAC)

influence mine A mine actuated by a force other than direct pressure or contact, such as sound or vibration (ORD)

INFO Prosign meaning information addressed (COMMS)

information Data that have not been evaluated as to their reliability or significance (INTEL)

information requirements (IR) Intelligence data on the enemy and/or environment that a commander needs in order to plan a course of action (INTEL)

infrared imagery/radiation Electromagnetic emissions at a frequency slightly below that of visible light (INTEL)

initial active duty training (IADT) Basic combat training and advanced individual (specialty) training administered to new soldiers (TNG)

initial issue The distribution of supplies or materiel to new soldiers or units, or when new equipment is available for distribution (LOG)

initial lane A cleared path through a minefield of minimum width (four meters) to allow passage of breaching and assault forces (TAC)

initial provisioning The process of planning the support for an item, including replacement parts and supplies, tools, test equipment, and the cataloging and distribution of these items (LOG)

initial radiation The burst of especially lethal radiation that occurs within the first minute after a nuclear detonation (NBC)

initiative Action that compels the enemy to concentrate on a defensive reaction rather than an offensive operation of their own (MS)

initiator A sensitive explosive that is used to detonate a larger quantity of a less sensitive explosive (ORD)

in kind Issue of actual quarters, food, or travel rather than a monetary allowance for them (LOG)

in place While maintaining formation or duty position (COLL)

in processing The administrative addition of a new soldier to a unit's rolls (ADMIN)

INSCOM United States Army Intelligence and Security Command (ADMIN)

inspection arms Combined command directing soldiers in formation to present their weapons for examination (DC)

inspector general (IG) The staff officer responsible for conducting inspections, investigations, surveys, and studies of a command's performance, discipline, morale, efficiency, and economy (ADMIN)

installation property Materiel that is authorized for siting at a specific facility rather than assignment to a unit or an individual (LOG)

installation type The classification of a CONUS installation according to its ability to support major units; in general, the classifications are as follows: Type A—corps-size, Type B—division-size, Type C—training center-size, Type D—Army or higher level or a variety of activities not included in types A, B, or C (ADMIN)

insurance stockage objective Inventory authorization based on an item's criticality or procurement lead time, not its recurring demand (LOG)

insurance type items Parts that are stocked even though there is no anticipated demand because their failure would affect the operation of a weapon system (LOG)

insurgency An organized revolutionary movement against a duly constituted government (UW)

INT Prosign meaning interrogative (COMMS)

integrated equipment Device that incorporates both communication and cryptocapability (COMMS)

integrated fire control system A system incorporating target acquisition, tracking, data computation, and engagement control functions (WPN)

integrated warfare The employment of conventional and nonconventional weapons in combination (MS)

integration The combination of recent and previously held intelligence to form the most up-to-date and accurate understanding possible (INTEL)

integrity Maintenance of a unit's doctrinal organization (MS)

intelligence Information that has been analyzed, evaluated, and refined to support the tactical or strategic decision-making process (ADMIN)

intelligence cycle The process by which intelligence needs are determined and fulfilled; direction, collection, processing, dissemination (INTEL)

intelligence estimate An appraisal, based on the evaluation of available information, of the enemy's probable courses of action (INTEL)

intelligence journal An official, permanent log of significant and/or classified events and messages occurring over a 24-hour period (INTEL)

intelligence requirement (IR) A commander's need for specific information on a given topic (INTEL)

intelligence summary (INTSUM) A frequent update of specific items of intelligence value (INTEL)

intensity factor A multiplier used to convert a standard day of supply into a combat day of supply; used in planning additional support needs for combat operations (LOG)

intercontinental ballistic missile Missile with a range of between 3,000 and 8,000 nautical miles (WPN)

interdiction The disruption of potential military power before it can be massed for use (TAC)

interdiction fire Rounds delivered with the objective of disrupting or denying the use of a specific enemy position (ARTY)

interdictive minefield Mine emplacements in enemy territory meant to disrupt, kill, and disorganize enemy operations (TAC)

interior guard Security force within an installation or command (TAC)

intermediate area illumination Lighting of the battlefield at the range of 2,000 to 10,000 meters (TAC)

intermediate maintenance Direct support maintenance to using units; calibration, repair, or replacement of components between user level and depot level (LOG)

intermediate marker A designated point of reference between an easily distinguished landmark and a minefield (TAC)

intermediate-range ballistic missile Missile with a range of between 1,500 and 3,000 nautical miles (WPN)

interment flag A cotton national flag measuring five feet by nine feet, six inches, that is used to drape the casket at a military funeral (DC)

intermittent arming Activation of mines only as required (ORD)

intermittent stream A stream that contains water on a cyclic or seasonal basis (NAV)

internal defense Measures taken by a government to protect itself and its citizens against subversion and insurgency (UW)

internal development The establishment and growth of institutions within a nation that serve the needs of its citizens (CA)

internal security A nation's prevailing state of law and order (STRAT)

international call sign A broadcast radio station's recognized identification code (COMMS)

international date line A line that approximately follows the meridian at 180 degrees longitude (in the Pacific opposite Greenwich, England) and divides the globe into separate calendar dates (ADMIN)

internee A person, during wartime, who is kept in any place against his or her will (LAW)

interoperability The capability of communications-electronics systems to exchange information (COMMS)

interpretability The quality of imagery—its suitability for providing usable information on a given target (INTEL)

interpretation Comparison of newly acquired information with previously developed intelligence and adjudgment of its significance (INTEL)

interrogation Systematic, direct questioning for the purpose of developing useful intelligence (INTEL)

interrupter Safety device that prevents the fuse from arming until a projectile has cleared the gun's muzzle (ORD)

intersection Method of estimating a geographic location by plotting the juncture of azimuths to it from two observing positions (NAV)

in the clear Unencrypted broadcast or message (COMMS)

INTREP Intelligence report (INTEL)

intrusion The intentional transmission of radio signals meant to confuse and/or deceive enemy communications (EW)

INTSUM Intelligence summary (INTEL)

IOE Irregular outer edge (ORD)

ionosphere Atmospheric layer that is especially useful for long-distance radio contact because of its reflection of sky waves (COMMS)

IP Initial point (TAC)

IPB **1.** Intelligence preparation of battlefield (INTEL); **2.** Intelligence property book (LOG)

IPR In-process review (LOG)

IR Infrared (EW)

Iroquois The UH-1H, or Huey, a light, rotary-wing, multipurpose aircraft (AVN)

IRR Individual Ready Reserve (ADMIN)

irregular forces Soldiers who are not official members of an army, are normally unpaid and untrained, and are armed only with their own weapons (MS)

irregular outer edge Short strips of mine clusters that extend from the turning point of baselines toward the enemy (TAC)

isoclinal A line on a map that connects points of equal magnetic variation (NAV)

isogonal A line on a map that connects points of equal magnetic declination (NAV)

isolation Preparation for an operation, during which troops are cut off from all contact other than their chain of command; promotes operations security and mission focus (TAC)

ITEP Individual training evaluation program (TNG)

J

J Joint (ADMIN)

J1 A joint staff personnel directorate (ADMIN)

J2 A joint staff intelligence directorate (ADMIN)

J3 A joint staff operations directorate (ADMIN)

J4 A joint staff logistics directorate (ADMIN)

J5 A joint staff plans and policy directorate (ADMIN)

J6 A joint staff communications electronics directorate (ADMIN)

J7 A joint staff civil-military operations directorate (ADMIN)

JAG Judge advocate general (LAW)

JAGC Judge Advocate General's Corps (ADMIN)

jamming The deliberate transmission or reflection of electromagnetic energy to interfere with enemy use of radio equipment or electronic devices (EW)

JAN grid Joint Army-Navy grid; a geographical coordinate system covering the whole earth while providing a degree of security over ordinary coordinate systems (NAV)

JATO unit Jet-assisted takeoff unit—a booster rocket (AVN)

JCS Joint chiefs of staff (ADMIN)

JD Julian date (ADMIN)

jet propulsion Motion induced by rapid, controlled burning of a fuel/air mix; *See* rocket propulsion (AVN)

JINTACCS Joint Interoperability of Tactical Command and Control Systems (COMMS)

JMPI Jumpmaster personnel inspection (ABN)

job book Pocket-sized listing of those tasks that make up enlisted specialties; contains spaces for recording individual soldiers' proficiency in the listed tasks (TNG)

Jody A song or rhythmic chant used to maintain coordinated movement of drilling personnel (TNG)

joint Combined operation of the services of a single nation, i.e., Army–Air Force (ADMIN)

joint airborne training Operational training that integrates the employment of air-delivered personnel and equipment with conventional ground operations (ABN)

joint chiefs of staff Those general military officers who head the U.S. Army, Navy, Air Force, and Marine Corps (ADMIN)

Joint Service Achievement Medal A decoration made in recognition of meritorious joint service (PERS)

Joint Service Commendation Medal A decoration made in recognition of meritorious achievement or valor in a combat area while part of a joint operation (PERS)

joint staff The advisors to a commander of a unified or specified command over two or more services (ADMIN)

J-TENS Joint Tactical Exploitation of National Systems (INTEL)

JTF Joint Task Force (ADMIN)

Judge Advocate General's Corps (JAGC) The combat service support branch that consists of attorneys and is responsible for legal administration (ADMIN)

Julian date calendar The sequential numbering of the days of the year, 1–366, without regard for months; facilitates planning; *See* Tables A-8 and A-9 (ADMIN)

jump The difference between the lines of departure and elevation as a round leaves the gun tube (ARTY)

jump boots Footgear with enhanced ankle support (PERS)

jumpmaster The soldier with command authority over all airborne personnel and equipment aboard an aircraft during an airdrop (ABN)

JUMPS Joint Uniform Military Pay System (ADMIN)

jump speed Airspeed suitable for parachutists to exit an aircraft safely (ABN)

jungle boots Lightweight footgear designed for easy drainage (PERS)

JUWTF Joint Unconventional Warfare Task Force (UW)

K Prosign meaning over, reply expected (COMMS)

Katusa Korean augmentee to United States Army (PERS)

KAWOL Knowledgeable, absent without leave; an AWOL soldier with knowledge of classified information (INTEL)

K-day The date for commencement of a convoy in support of a particular operation; a planning point within the wartime manpower planning system (ADMIN)

Kevlar® A projectile-resistant fabric (LOG)

key **1.** A map's legend (COLL); **2.** A data listing or device used to encrypt or decrypt a message (COMMS)

keyholing Tumbling of a bullet in flight due to lack of spin on the round (WPN)

key item An item of materiel stored at only one point (LOG)

key list A directory showing the cryptosystem in use for given crypto-period (COMMS)

key position A federal civilian employee not subject to military activation because his or her civilian job is important to national security (ADMIN)

key symbol In psychological warfare, any element (e.g., design, music, or color) with a powerful cultural significance to a target audience (UW)

key terrain Any area, the holding of which affords a significant advantage to either combatant (TAC)

KGB The Soviet Union's Committee for State Security, an intelligence and border security organization (OPFOR)

khaki Sand-colored durable cotton cloth frequently used for uniforms (LOG)

KHz Kilohertz (COMMS)

KIA Killed in action (ADMIN)

killing zone A sector to which a commander plans to force his or her enemy in order to bring maximum firepower against enemy targets (TAC)

kiloton A measure of explosive power equivalent to 1,000 tons of TNT (WPN)

kinetic energy ammunition Projectile whose destructive force derives from its powerful motion upon impact rather than its explosive effect (ORD)

Kiowa The OH-58C helicopter, a light observation, rotary-wing aircraft (AVN)

kitchen police (KP) **1.** Non–mess hall personnel temporarily assigned to assist in a dining facility (COLL); **2.** Those who perform any food preparation or dining facility maintenance duties other than actual cooking (LOG)

K-kill Complete kill; destruction of a vehicle or weapons system and its crew to the point where it cannot perform its mission (ORD)

km Kilometer (NAV)

knife rest A portable frame on which barbed wire is strung for use as a temporary barricade (TAC)

KP Kitchen police (COLL/LOG)

KT Kiloton (NBC)

L Symbol for the blister agent lewisite (NBC)

labeled cargo Materiel in transit that is of a hazardous nature, identified by the following color codes: green—pressurized gas; red—flammable items; white—acids and corrosives; and yellow—flammables, oxidizers (LOG)

Lance A surface-to-surface guided missile designed to support at corps level with long-range fire (WPN)

lands Raised surface between the rifling in the bore of a gun (WPN)

land tail The portion of an airmobile unit that moves via surface transportation (ADMIN)

lane A clear route through an obstacle (TAC)

LAPES Low-altitude parachute extraction system (ABN)

large-scale map A map at a scale of 1:75,000 or larger (INTEL)

laser designator A device that uses a beam of laser energy to mark an object (WPN)

laser-guided A weapon targeting system that homes on reflected laser energy (WPN)

laser seeker In a laser-guided system, the device that detects the reflected energy and homes the weapon on the target (WPN)

laser target designation The holding of a laser beam on a target to permit a projectile to home on it (WPN)

lashings Tie downs that prevent the shifting of materiel during shipment (LOG)

LASINT Laser intelligence (INTEL)

last four Final digits of a Social Security number; used as an identification number within small groups (ADMIN)

lateral Trench or underground gallery cut parallel to the front (ENG)

lateral route A road roughly parallel to the forward edge of the battle area (TAC)

lateral transfer The shift of materiel from a unit with an excess to a unit with a requirement for it (LOG)

latitude Distance north or south of the equator; measured in degrees (NAV)

latrine A toilet facility (LOG)

launcher Any structure designed to support a missile for firing (AD)

LAW Light antitank weapon (WPN)

laws of war Internationally recognized agreements that describe the limits of conduct during combat between nations (LAW)

lay, laying To emplace or aim a large gun (ARTY)

LBE Load-bearing equipment (INF)

LC **1.** Loud and clear (COMMS); **2.** Line of contact (TAC)

LCC Logistic control code (LOG)

LD Line of departure (TAC)

LD/LC Line of departure/line of contact (TAC)

lead Point of aim ahead of a moving target (ARTY)

leaders' reaction course Course that tests individuals' ability to motivate small groups to work together to overcome difficult obstacles (TNG)

leave Authorized absence from duty, earned at the rate of 1.5 days per month and deducted from the individual's leave account when used (PERS)

leave en route Ordinary leave taken in addition to allowed travel time between duty assignments (PERS)

left face A two-part command (preparatory: left; execution: face) directing troops in formation to execute a 90-degree turn to the left (DC)

leg Infantryman—one who travels on foot; also, any nonairborne soldier (COLL)

legal assistance program A program to aid soldiers and their families with personal, noncriminal legal difficulties (LAW)

Legion of Merit Decoration made to key individuals for exceptionally meritorious conduct of outstanding services; recipients may be military or civilian, American citizens or foreign nationals (PERS)

lensatic compass Compass fitted with a magnifying lens to aid in the accuracy of reading the azimuth scale (NAV)

LES Leave and earnings statement (ADMIN)

LET **1.** Live environment training (TNG); **2.** Light equipment transporter (LOG)

levee en masse Concept under which noncombatant civilians may be treated as regular combatants if they join in of the battle (LAW)

level of protection The degree of preservation provided by the packaging of an item of supply, broken down as follows: Level A—military protection, Level B—limited military protection, and Level C—minimum military protection (LOG)

LF Low frequency (COMMS)

liaison The communication among commands necessary to maintain mutual understanding and unity of action (ADMIN)

liberation theology The use of religious symbolism in propaganda attempting to justify revolutionary activity (UW)

LIC Low-intensity conflict (UW)

lieutenant **1.** A junior commissioned officer whose rank is either first lieutenant (pay grade O-2) or second lieutenant (pay grade O-1) (ADMIN); **2.** An assistant (COLL)

lieutenant colonel A commissioned officer; pay grade O-5 (ADMIN)

lieutenant general A commissioned officer; pay grade O-9 (ADMIN)

lifer A career soldier, one committed to the Army at least through eligibility for retirement (COLL)

light Descriptive of a force with little or no armor assets (COLL)

light damage A degree of destruction that does not prevent the immediate use of equipment or facilities; only minor repairs are required (TAC)

light discipline Nighttime restrictions on lighting to minimize enemy surveillance (TAC)

light duty Temporary excusal from strenuous physical activity due to illness or injury (MED)

light level of operations Combat employing less than 30 percent of maneuver echelons and 50 percent of fire support assets in sporadic action, with no requirements for support from higher echelons (TAC)

light line A line forward of which vehicles can use only blackout lights at night (TAC)

light shelter A defensive position capable of withstanding continuous bombardment by eight-inch shells (TAC)

limited standard type Items that do not meet Army operational specifications but are retained for training purposes or for other requirements (LOG)

limited storage Storage for 90 days or less—the least protected category of storage (LOG)

LIN Line item number (LOG)

linear speed method Means of determining a moving target's future position—multiplication of its speed by the estimated time of the projectile's flight (ARTY)

line crosser A combatant who surrenders to the enemy while not in immediate danger (LAW)

line item number (LIN) An alphanumeric designator for a specific supply requirement to be filled by very similar but not necessarily identical items (LOG)

line of bearing The angular measure in degrees from true north from a given point to the source of a radio signal (EW)

line of contact A tracing of the points along which opposing forces are engaged (TAC)

line of departure **1.** Line along which troops are deployed prior to a coordinated movement or attack (TAC); **2.** A line tangent to a round's trajectory as it leaves the tube (ARTY)

line of drift Natural path along which wounded soldiers can be expected to withdraw for medical assistance (MED)

line of duty **1.** A medical disability not resulting from fault or neglect of the soldier concerned (MED); **2.** Within one's range of authorized responsibilities (ADMIN)

line of elevation The extended axis of a laid gun's tube (ARTY)

line officer A combat-branch officer of a line unit (PERS)

line-route map A chart of the locations of communications circuitry, switchboards, and stations (COMMS)

line search Reconnaissance of a specific line of communications (INTEL)

lines of communications Air, land, and water transportation routes among operational bases (TAC)

line unit A combat element; a frontline unit (TAC)

link A single, direct communications channel between two terminals (COMMS)

link callsign A combination of characters or words that identifies a communications channel between two terminals (COMMS)

listening silence A command-ordered period during which net operating frequencies are to be monitored but only emergency transmissions broadcast (COMMS)

Lister bag A canvas bag for the storage, evaporative cooling, and dispensing of water to troops in semifixed field exercises (TAC)

litter Any device used to move injured personnel; a stretcher (MED)

live cluster A semicircular emplacement of one to five mines within a two-meter radius (TAC)

LO **1.** Lubrication order (MAINT); **2.** Liaison officer (ADMIN)

loading plan The detailed instructions for the arrangement of personnel and equipment being transported (LOG)

LOB Line of bearing (EW)

LOC Lines of communication (TAC)

local purchase The authorization to acquire an item with government funds, outside the centralized purchase system (LOG)

locked on Status of a target-seeking system that is continuously and automatically following its target (WPN)

lodgement A foothold that has been gained in formerly enemy territory (TAC)

LOGEX Logistical exercise (TNG)

logistics The movement and maintenance of military forces, including acquisition and distribution of materiel, transport and feeding of soldiers, provision of facilities, and furnishing of other miscellaneous support services (MS)

logistics immaterial position A duty position that can be filled only by a commissioned officer with an ordnance, quartermaster, or transportation specialty (ADMIN)

logistic support Provision of sufficient materiel and services to a maneuver element to enable it to accomplish its mission (TAC)

LOGMARS Logistics application of automated marking and reading symbology (LOG)

LOI Letter of instruction (TNG)

LOMAD Low-to-medium-altitude air defense (AD)

longitude The east-west position of a point on the earth's surface; measured by arcs of the equator (NAV)

long-life item Materiel with an expected service life of more than 20 years (LOG)

long-range radar Target-detection equipment capable of resolving a one-square-meter target at a range of 300 to 800 miles (INTEL)

long supply Materiel on hand in a service's inventory in excess of M-day requirements (LOG)

long-tour area A locale overseas where duty assignments are 36 months (with dependents) or 24 months (without dependents) or longer (ADMIN)

loop sling Means of achieving steady hold and aim of a rifle by wrapping the sling snugly around the non-firing arm (TAC)

LOS Line of sight (TAC)

lost Report by a forward observer that a round's impact was not seen (ARTY)

lot number A control number identifying a group of identical supply items that were manufactured together (LOG)

lowering line Rope or other attachment between a parachutist and his or her rucksack; upon opening of the parachute, the jumper lowers his or her equipment on this line (ABN)

low explosive Material that changes from a solid to a gaseous state at a destructive rate of less than 400 meters per second; used in igniter trains and as a propellant (ORD)

low-grade cryptosystem An encryption method meant to provide expedient security for tactical communications (COMMS)

low-intensity conflict The limited application of force for political purposes by nations or organizations to coerce, control, or defend a territory, or to establish or defend rights; includes operations by or against irregular forces, peace-keeping operations, and terrorism and counterterrorism (UW)

low-level flight Minimum-altitude aviation operations that follow the earth's contours to minimize the visibility of the aircraft and evade ground fire (AVN)

low nuclear yield Nuclear weapon burst that generates 1 to 10 kilotons of blast energy (WPN)

low-order burst The fragmentation of a projectile into a few large pieces rather than many small pieces (ORD)

low quarters Dress shoes worn with a class A uniform (COLL)

low velocity A projectile muzzle velocity of 2,499 feet per second or less (ARTY)

low-velocity drop Air delivery in which the drop velocity does not exceed 30 feet per second (ABN)

low-visibility operations A maneuver or mission that, because of its sensitive nature, is conducted so as not to draw more attention than necessary; clandestine or covert options are deemed unnecessary or not feasible (TAC)

LP Listening post (TAC)

LRSU Long-range surveillance unit (TAC)

LTC Lieutenant colonel (PERS)

LTG Lieutenant general (PERS)

lubrication order (LO) Mandatory guidance on proper preventive application of lubricants to certain machinery (MAINT)

LZ Landing zone (AVN)

M Military (ADMIN)

M60 An air-cooled, belt-fed, gas-operated, light automatic weapon; caliber 7.62 mm (WPN)

M88 A tracked vehicle able to tow and recover tanks (LOG)

MAAG Military assistance advisory group (ADMIN)

MAB Mobile assault bridge (ENG)

MAC Military airlift command (LOG)

mack stem The shock wave formed by the combined incident (airburst) wave and the reflection of the incident wave by the ground resulting from a nuclear detonation (NBC)

MACO Marshaling area control officer (ABN)

macom Major Army command (ADMIN)

MAG Military Advisory Group (ADMIN)

magazine **1.** A holding clip used to automatically feed separate rounds to a weapon (WPN); **2.** A storage area, especially for munitions (LOG)

magnetic azimuth A direction determination based from magnetic north (NAV)

magnetic bearing The direction to a point, expressed as an angle measured clockwise from magnetic north (NAV)

magnetic declination The angular difference between magnetic and geographical meridian at any particular point on the earth's surface (NAV)

magnetic north The direction sought by the magnetic needle of a compass (NAV)

mail cover Authorized surveillance of personal correspondence (INTEL)

main defense area The sector from the forward edge of the battle area to the rear boundary of the command's defense forces (TAC)

main line of resistance A line at the forward edge of the battle area along which mutually supporting fires are coordinated (TAC)

maintain watch Continuous monitoring of a given radio frequency (EW)

maintenance concept The plan for support and repair of an end item (LOG)

maintenance float Mission essential equipment authorized in excess of operational needs to substitute for like units during their scheduled or unscheduled repair (LOG)

MAIT Maintenance assistance and instruction team (LOG)

MAJ Major (PERS)

major Army command A command directly subordinate to Headquarters, Department of the Army (ADMIN)

major general A commissioned officer; pay grade O-9 (PERS)

major inventory variance A discrepancy of $200 or more in the dollar value of a stock item (LOG)

Major NATO Commanders The Supreme Allied Commander Atlantic, Supreme Allied Commander Europe, and Allied Commander in Chief Channel (STRAT)

major nuclear power A nation capable of striking all other nations with a nuclear weapon (STRAT)

major weapon system A system designated by the Department of Defense as vital to the national interest (LOG)

make safe To disarm or disable; to make incapable of operation (WPN)

management The establishment of objectives and controlled utilization of resources in fulfillment of responsibilities (ADMIN)

management control number Number assigned to certain items of materiel to facilitate their close supervision (LOG)

mandatory recoverable item An item of materiel that is not consumed in use and must be returned for repair or reconditioning (LOG)

mandrel A mold used in forming cartridge cases or the solid propellants for rockets (ORD)

maneuver Movement on the battlefield to attain an advantage over the enemy (TAC)

maneuvering force Combat element that seeks to achieve an objective by moving to a more advantageous position with respect to the enemy (TAC)

manifest A detailed listing of passengers or equipment being transported (LOG)

manipulative communications cover (MCC) Measures taken to limit an enemy's ability to identify or detect friendly communications (EW)

manipulative communications deception (MCD) The alteration or simulation of friendly radio signals for the purpose of misleading the enemy (EW)

manipulative electronic deception The alteration of friendly electromagnetic emissions to eliminate information that could be of value to the enemy (EW)

manning level A personnel ceiling imposed against the normal authorization because of a lack of resources (ADMIN)

manning table A listing of a unit's soldiers and the duty positions to which they are assigned (ADMIN)

MANPAD Man-portable air defense (AD)

man-portable Capable of being carried by one soldier in addition to his or her personal equipment; normally weighing no more than about 30 pounds (LOG)

manual of arms Drill performed with a weapon for the purpose of bolstering pride and concentration (DC)

map exercise (MAPEX) A training situation limited to the maneuver of notional units to utilize the terrain on a given map (TNG)

map K A factor that corrects for the discrepancy between the scales of the firing chart and the plotting scale in use (ARTY)

map orientation The alignment of a map's directions with real ground directions to enhance the users' sense of their position (NAV)

map reconnaissance Examination of terrain in preparation for operations by use of a map instead of actual observation of the terrain (INTEL)

map scale The ratio between map distance and ground distance; for example, if the map scale is 1:50,000, one map inch represents 50,000 ground inches (NAV)

map series An area grouping of maps having the same scale and specifications (NAV)

march unit Element that moves under the control of a single commander (TAC)

marginal weather Weather bad enough to impose procedural limitations on operations (TAC)

mark A command of execution in any operation requiring precise synchronization (ADMIN)

marking fire Fire delivered on a target for the purpose of identifying it (ARTY)

marking panel A large sheet of material used for visual communication between the air and the ground (COMMS)

marking team An advance party whose mission is to establish landing and/or drop zones through use of markers and navigational beacons (TAC)

marksman The minimum qualification level for accurate delivery of fire with an aimed weapon; below sharpshooter (PERS)

mark time **1.** To march in place (DC); **2.** To wait (COLL)

marmite can An insulated food container (LOG)

MARS Military Affiliate Radio System (COMMS)

marshaling Mobilization in one place for embarkation (TAC)

marshaling area A site where troops are assembled, maintained, and organized for embarkation (ADMIN)

martial law The exercise of governmental authority over civilians by military officials (CA)

MASH Mobile army surgical hospital (MED)

MASINT Measurement and signature intelligence (INTEL)

masking The use of decoy transmitters to obscure the operation or location of a critical transmitter (EW)

mass Concentrated combat power (TAC)

mass casualties A quantity of killed or wounded soldiers that overwhelms the capacity of available medical support and jeopardizes the successful completion of the mission (MED)

mass formation An assembly of company size or larger in which the squads in a column are abreast of each other (DC)

master depot Theater-level control point for entire classes of supply or designated items (LOG)

master force list A status report on each requirement of a given operation plan (ADMIN)

master menu A monthly supply bulletin listing Army-wide standard meals and menu planning guidance (LOG)

master sergeant An enlisted noncommissioned officer; pay grade E-8 (PERS)

MATCAT Materiel category (LOG)

MATES Mobilization and training equipment sites (LOG)

maximum effective range The maximum range at which a weapon is designed to accurately deliver its destructive force (WPN)

maximum gradability The steepest degree of slope that a vehicle is able to negotiate in low gear (LOG)

maximum hospital benefit The point of progress during a patient's hospitalization when continued hospitalization is not required to maintain recovery (MED)

maximum ordinate The high point in the trajectory of a projectile (ARTY)

maximum range Greatest distance at which a weapon can deliver fire, without consideration of accuracy (WPN)

mayday International code word to call for assistance in life preservation (ADMIN)

MBA Main battle area (TAC)

MBT Main battle tank (ARM)

MCD Manipulative communication deception (EW)

MCI Meal, combat, individual (LOG)

MCSR Material condition status report (LOG)

MD Symbol for the blister agent methyldichloroarsine (NBC)

M-day The date on which mobilization for an operation commences; a planning point within the wartime manpower planning system (ADMIN)

M-day force Personnel strength available on the first day of mobilization (ADMIN)

meaconing The broadcast or rebroadcast of false navigational signals for the purpose of disrupting air traffic (EW)

meal surcharge Money that is collected from certain personnel authorized to subsist at a dining facility and is applied toward operating costs (LOG)

measle chart Any map or map overlay depicting the position of numerous, similar positions (COLL)

measurement ton 40 cubic feet—a unit of measure for materiel movement aboard ships (LOG)

mechanized A force capable of movement by motorized vehicles (ADMIN)

MED Manipulative electronic deception (EW)

Medal of Honor Decoration for conspicuous heroism in combat at the risk of life; failure to perform the act would not subject the soldier to censure for failure to perform his or her duty (PERS)

MEDCOM Medical command (MED)

medevac Medical evacuation (MED)

median lethal dose The level of exposure to chemicals or nuclear radiation that would be fatal to 50 percent of exposed personnel (NBC)

medical intelligence The assessment of information regarding the capability of enemy forces to conserve their fighting strength and the impact of that capability on friendly strength (INTEL)

medium nuclear yield Nuclear weapon burst that generates 10–50 kilotons of blast energy (WPN)

medium-range ballistic missile A missile with an effective range of 600 to 1,500 miles (WPN)

medium-range radar Target-detection equipment capable of distinguishing a one-square-meter target at a range of 150 to 300 miles (INTEL)

medium-scale map A map at a scale of between 1:75,000 and 1:600,000 (INTEL)

meeting engagement Combat resulting when a maneuvering force unexpectedly contacts the enemy (TAC)

megaton A measurement of the explosive power of a nuclear weapon; equivalent to the energy released by the explosion of one million tons of TNT (WPN)

MEPS Military Entrance Processing Stations (ADMIN)

mercenary A soldier for hire; one not necessarily motivated by political loyalty (PERS)

Meritorious Service Medal Decoration made in recognition of outstanding noncombat achievement or service (PERS)

mess Related to dining or kitchen operations (LOG)

message Ideas or data prepared for transmission in either plain or encrypted language (COMMS)

message indicator A set of symbols in an encrypted message that describes the arrangement of its cryptovariables (COMMS)

message parts Apparently unrelated segments of a message that has been separated for additional security (COMMS)

mess kit Utensils for eating in the field (PERS)

mess kit repair company Mythical unit whose mission or location is considered obscure by the speaker (COLL)

MET Medium equipment transporter (LOG)

method of resupply Means of communicating supply needs from a unit to an issuing facility, i.e., requisition, automatic, usage report, etc. (LOG)

METL Mission essential task list (TNG)

METT-T Mission, Enemy, Terrain, Troops—Time available (TAC)

MFF Military free fall (ABN)

MFL Master force list (ADMIN)

MFT Master fitness trainer (TNG)

MG Major general (PERS)

MHz Megahertz (COMMS)

MI Military intelligence (ADMIN)

MIA Missing in action (ADMIN)

MICLIC Mine-clearing line charge (ORD)

MICOM Missile command (ADMIN)

MIJI Meaconing, intrusion, jamming, and interference (EW)

MILES Multiple integrated laser engagement simulation; a system for simulating fire and its effects on targets during training exercises (TNG)

Military Affiliate Radio System A network of amateur radio operators that serves as an emergency auxiliary to normal communications channels (COMMS)

Military Airlift Command (MAC) The Air Force agency that manages airlift capability in support of Army operations (LOG)

military area A specific geographical region of such military significance that it is declared such by authority of the President or the secretary of defense (ADMIN)

military assistance advisory group (MAAG) The control structure for delivery of counsel and training to the forces of a foreign host nation (ADMIN)

military capability The ability to accomplish a wartime objective; measured through four components—force structure, modernization, readiness, and sustainability (MS)

military civic action The employment of military personnel in public service activities, such as education, agriculture, sanitation, and social development (UW)

military commission A court convened by military authority for trial of nonmilitary personnel accused of violation of the laws of war or the military government (LAW)

military courtesy Traditional social conduct practiced among military personnel; typified by the hand salute (ADMIN)

military crest A line or point below the maximum altitude of a hill or ridge line that permits observation or movement without the increased exposure of the actual crest (TAC)

military grid reference system Position referencing and direction computation based on the standard grid squares depicted on military maps (NAV)

military intelligence Information that is gathered, refined, and evaluated in support of the planning and conduct of military operations (ADMIN)

Military Intelligence Corps The combat support branch and personnel responsible for the collection, processing, and dissemination of militarily significant information (ADMIN)

military jurisdiction The authority to impose military law rather than local civilian law (LAW)

military load classification A system of rating the load-bearing capacity of roads, bridges, and rafts (LOG)

military necessity The right of belligerents to apply any measures necessary, within the laws of war, to accomplish their military objective (MS)

military nuclear power A nation with nuclear weapons and the means to use them (STRAT)

military occupational specialty (MOS) A group of related skills that constitute an individual soldier's job (TNG)

military pay order Form used to change a soldier's pay status (ADMIN)

Military Police Corps Combat support branch and personnel responsible for enforcing the Uniform Code of Military Justice, military law over civilians, and treatment of enemy prisoners of war (ADMIN)

mill A keyboard for transcription of intercepted communications (EW)

MILPERCEN Military Personnel Center (ADMIN)

MILPO Military Personnel Office (ADMIN)

MILSTAMP Military standard transportation and movement procedures (LOG)

MILSTRIP Military standard requisitioning and issue procedures (LOG)

mine A stationary bomb designed to damage troops or vehicles that move within its range (ORD)

minefield lane An unmined path through a minefield (TAC)

mine strip Two parallel rows of emplaced clusters offset three meters to either side of a centerline (TAC)

mine sweeper A roller attached to the front of an armored vehicle for the purpose of detonating pressure-sensitive mines (ENG)

minimize Order to suppress routine message traffic to keep available channels open for anticipated priority traffic (COMMS)

minimum range Nearest distance that a gun can fire on without harming friendly forces or itself (WPN)

mining system Tunneling to reach under enemy positions, either to provide access for troops or to emplace explosive charges (ENG)

miosis Inability of the eyes to dilate properly due to absorption of small amounts of nerve agents (MED)

misfire The failure of a weapon to launch its round or of a munition to detonate as planned (WPN)

missile A self-propelled projectile (WPN); *See also* prepared missile, ready missile

missile effective rate The percentage of tactical missiles that are either prepared or ready missiles (WPN)

missile monitor A mobile, electronic system by which air defense commanders observe and control their fire distribution (AD)

mission A stated military objective along with guidance on how it is to be executed (TAC)

mission capable equipment Equipment that is in good repair and able to perform the function for which it is intended (MAINT)

mission load That quantity of class II (clothing/tools) and class IV (construction material) supplies authorized to be on hand to support a unit's peacetime and combat mission until resupply can be effected (LOG)

mission support site A relatively secure base for storage and operations staging (UW)

mission type order A duty assignment that states objectives without specifying how they are to be attained (ADMIN)

mixed An observer's report of an approximately equal number of air- and groundbursts (ARTY)

mixed forest A growth of deciduous and coniferous trees covering 25 percent to 75 percent of a sector of land and having a canopy height of up to 20 meters (TAC)

mixed medical commission A multinational board of doctors who evaluate sick and wounded prisoners of war (MED)

mixed minefield A minefield made up of both antitank and antipersonnel mines (TAC)

mixed salvo A series of rounds, some of which fall short and some of which go beyond a target (ARTY)

M-kill Mobility-kill; damage sufficient to immobilize an armored vehicle but the vehicle is repairable; crew and weapons still function (ORD)

MLRS Multiple launch rocket system (ARTY)

MMC Materiel management center (LOG)

MOA **1.** Memorandum of agreement (ADMIN); **2.** Military operation area (STRAT)

MOBDES Mobilization designee (ADMIN)

mobile defense Maneuver to make maximum use of the terrain to hold the initiative, even though not attacking (TAC)

mobile training team An expert element detailed to assist another command or nation in learning to accomplish a mission or procedure (TNG)

mobile unit An element with sufficient transportation capability to move itself in one trip (LOG)

mobile warfare Conflict in which the opposing forces seek to maintain the initiative by maneuver and utilization of key terrain (TAC)

mobility The ability of a unit to move while retaining the ability to perform its primary mission (MS)

mobilization The assembly and organization of resources in support of military objectives (STRAT)

mobilization base units Those Reserve elements preselected for activation upon initial mobilization (ADMIN)

mobilization designee A member of the Individual Ready Reserve trained to occupy a vacant active duty position during early mobilization (ADMIN)

mode of transport Basic medium for movement—air, land, or sea (LOG)

Moderate damage A degree of destruction that prevents the use of equipment or facilities until extensive repairs are made (TAC)

moderate level of operations Combat employing 30 percent to 60 percent of available maneuver echelons and over 50 percent of available fire support over a period during which support from higher echelons to achieve the mission is not likely to be needed (TAC)

moderate risk A degree of anticipated danger from the effects of nuclear weapons; where expected casualties would be minor, not affecting a unit's combat efficiency (NBC)

modification work order A written requirement and instructions to adjust or change an item of equipment (LOG)

modified resection A method of determining one's exact position when located along a linear terrain feature and when an identifiable terrain feature of known location is visible in the distance; a magnetic azimuth to that landmark is taken and converted successively to a back azimuth and grid azimuth that is then plotted—the intersection with the original linear terrain feature determines position (NAV)

modulation The process of varying a carrier radio wave in order to transmit information (COMMS)

MOGAS Motor gasoline (LOG)

Mohawk The OV-1, a fixed-wing observation, electronic warfare, or lightly armed aircraft (AVN)

Molotov cocktail Field-expedient antiarmor charge made of a breakable container filled with a gas/oil mixture and stopped with a cloth wick that is lit prior to throwing the charge (ORD)

monitoring The detection of, and continued listening to, a signal (COMMS)

MOPMS Modular pack mine system (ORD)

MOPP Mission-oriented protection posture (NBC)

morale support activities A post branch whose mission is to provide recreation, craft, and hobby-type activities (ADMIN)

morning report The daily summary of the personnel status of a unit (ADMIN)

Morse code A system of transmitting characters by varying combinations of long and short impulses; *See* Table A-2 (COMMS)

mortar A short-range, muzzle-loaded indirect fire cannon (WPN)

MOS Military occupational specialty (TNG)

motorized A unit equipped with the ability to transport itself via vehicles that are not necessarily armored (ADMIN)

motor march Controlled movement of vehicle-mounted troops (TAC)

motor pool A reserve of vehicles available for common use as needed (LOG)

motor stables Preventive and routine maintenance performed on vehicles at the unit level (MAINT)

mounted In or on a vehicle, as opposed to on foot (COLL)

mousetrap The M5, a mine-firing device activated by pressure release (ORD)

MOUT Military operations on urbanized terrain (TAC)

movement control The planning and regulation of the flow of traffic along lines of communication (LOG)

movement priority designator A rating of the precedence of materiel when allocating limited cargo-moving capacity (LOG)

movement to contact The approach to a position from which the enemy can be engaged (TAC)

moving screen A patrol whose mission is to keep enemy scout elements at a sufficient distance from the main body to prevent observation (TAC)

moving target indicator A radar feature that cancels display of stationary objects (INTEL)

MP Military police (ADMIN)

MPI **1.** Military police investigator (ADMIN); **2.** Mean point of impact (ARTY)

MPL Mandatory parts list (LOG)

MPRJ Military Personnel Records Jacket (ADMIN)

MRB Motorized rifle battalion (OPFOR)

MRD Mandatory retirement date (ADMIN)

MRE Meal, ready to eat (LOG)

MRO Materiel release order (LOG)

MRR Motorized rifle regiment (OPFOR)

MSC Medical service corps (ADMIN)

MSG Master sergeant (PERS)

MSGID Message identifier (COMMS)

MSM Meritorious Service Medal (PERS)

MSR Main supply route (LOG)

MTLR Moving target locating radars (ARTY)

MTMC Military traffic management command (LOG)

MTOE Modification table of organization and equipment (ADMIN)

MTT Mobile training team (TNG)

mufti Civilian clothing (PERS)

multichannel The transmission of two or more voice-grade channels over a single carrier frequency (COMMS)

multiple unit training assemblies The four-hour blocks of training under which Reserve units meet (TNG)

multiple warning Reliance on more than a single type of sensing/warning system in order to attain the most credible information possible (INTEL)

multiplexing Simultaneous transmission of more than one channel of information; permits two-way communication (COMMS)

multisection charge A propelling charge that permits range adjustment by varying the number of powder bags (ARTY)

munitions Military supplies, particularly weapons and ammunition (LOG)

MUSARC Major United States Army Reserve Command (ADMIN)

MUST Medical unit, self-contained, transportable (MED)

mustard gas A liquid, chemical blister agent (NBC)

muster Assembly of all of a unit's personnel or equipment for accountability purposes (ADMIN)

MUTA Multiple Unit Training Assembly (TNG)

mutual support The ability of units to assist each other based on their assets, abilities, and relative positions (TAC)

MUX Multiplex (COMMS)

muzzle The open end of a gun's barrel (WPN)

muzzle brake A vent near the open end of a gun's barrel that permits the escape of explosive gases to reduce recoil (ARTY)

muzzle burst Premature explosion of a round as it leaves the barrel of a weapon (ARTY)

muzzle compensator A device designed to control muzzle movement through the controlled venting of gas (WPN)

muzzle velocity The speed of a projectile as it leaves a gun's barrel (WPN)

MWO Modification work order (LOG)

NA Not applicable (ADMIN)

NAAK Nerve agent antidote kit (NBC)

NAC National agency check (INTEL)

NAI Named Area of Interest (INTEL)

napalm (NP) Gelatinized gasoline or oil used in bombs and flamethrowers (ORD)

nap-of-the-earth flight Low-level flight that follows the terrain's contours to minimize visibility and evade ground fire (AV)

NARCINT Narcotics intelligence (INTEL)

national agency check (NAC) Basic background check of federal agencies' records regarding applicants for security clearances (INTEL)

National Command Authorities The highest U.S. military commanders—the President, the Secretary of Defense, and their designated alternates or successors (ADMIN)

national defense area Temporary designation of non-Federal land for DOD control, normally for the protection of classified or sensitive equipment or information (ADMIN)

national intelligence Evaluated information whose scope and importance has broad impact on national security, beyond the concern of a single department or agency (INTEL)

national intelligence estimate (NIE) A strategic-level composite of friendly intelligence assessments of the intentions, capabilities, and vulnerabilities of foreign nations (INTEL)

national inventory control point A distribution and management center for an item or type of materiel (LOG)

national item identification number (NIIN) A nine-digit component of a national stock number consisting of a country-of-origin designation and a serial number (LOG)

National Security Agency (NSA) Agency responsible for U.S. signals intelligence (INTEL)

National Security Council (NSC) The board of senior advisors whose mission is to counsel the President and assist him in integrating domestic, foreign, and military policies in support of national defense (STRAT)

National Security Organization The umbrella grouping of agencies responsible for the national defense—the Commander in Chief, National Security Council, Department of Defense, Central Intelligence Agency, and Office of Emergency Planning (STRAT)

national stock number (NSN) A control number for any item of materiel in the supply system; consists of a four-digit federal supply classification code and a nine-digit national item identification number (LOG)

National War College The senior-level training school for all services, offering courses in strategic studies and national security; located at Fort Leavenworth, Kansas (TNG)

NATO North Atlantic Treaty Organization; a military alliance in support of the collective security of the United States and various nations of western Europe (STRAT)

NATO earmarked forces International military forces designated to be placed under the operational control of NATO under given contingencies (STRAT)

nautical mile 6,076.1 feet or one minute of arc on the earth's surface (NAV)

nautical twilight That time of incomplete darkness when the center of the sun is 12 degrees below the horizon (ADMIN)

NBC Nuclear, biological, chemical (ADMIN)

NCA National Command Authority (ADMIN)

NCO Noncommissioned officer (PERS)

NCOES The Noncommissioned Officer Education System (TNG)

NCS Net control station (COMS)

near real time Delay between the occurrence of an event and its remote observation due to delays introduced by transmission, observation, or data processing equipment (INTEL)

need to know The restriction of certain classified information to individuals with an operational requirement for knowledge of it (INTEL)

negligible risk A degree of anticipated danger from the effects of nuclear weapons; personnel are thought to be safe, with only some temporary damage to vision probable (NBC)

nerve agent A chemical weapon that paralyzes and kills through interference with nerve impulses (NBC)

NESTOR A family of secure communications devices (COMMS)

NET New equipment training (TNG)

net A communications structure controlled by and reporting to a single command center (COMMS)

net call sign A signal that alerts all stations within a communications structure (COMMS)

net control station (NCS) Radio terminal that directs traffic and enforces discipline within a communications structure (COMMS)

neutrality The impartiality of a nation toward nations engaged in hostilities against each other (LAW)

neutralization fire Fire that knocks the enemy out of action for a limited time (TAC)

neutral persons Nationals of a state not taking part in the war (LAW)

NFL No fire line (TAC)

NG National Guard (ADMIN)

NICP National inventory control point (LOG)

night traffic line Points beyond which wheeled vehicles cannot proceed during hours of darkness (TAC)

NIIRS National image interpretability rating scale (INT)

NLT No later than (COLL)

NMC Not mission capable (LOG)

NMI No middle initial (ADMIN)

NNNN Prosign meaning no, incorrect (COMMS)

NOD Night observation device (WPN)

node A point where information from two or more links or systems converges (COMMS)

NOE Nap of the earth (AVN)

no-fire line A line, short of which artillery may not deliver rounds and beyond which they may fire without danger to friendly forces (TAC)

no go **1.** Failure of an evaluation (TNG); **2.** Equipment status—unserviceable (LOG); **3.** Terrain that is impassable for vehicles (TAC)

noise Any unwanted receiver response other than desired radio signal reception (COMMS)

noise discipline The suppression of all unnecessary sound to minimize the chance of detection by the enemy (TAC)

NOK Next of kin (ADMIN)

nomenclature The official, detailed, and standard name assigned to an item in the supply system (LOG)

Nomex A fire-resistant fabric (LOG)

nominal weapon A nuclear weapon yielding approximately 20 kilotons (ORD)

nonappropriated funds Money raised through the efforts of civilian and military personnel activities to enhance the military community without having to rely on government funding (ADMIN)

nonboresafe fuze Fuze that lacks a device that prevents its possible detonation before leaving the gun's bore (ORD)

noncommissioned officer (NCO) An enlisted soldier in pay grades E-4 through E-9 who functions as a first line trainer and supervisor of other soldiers; also called a noncom (PERS)

nonduty status A soldier who is unavailable for duty because of confinement, hospitalization, leave, being AWOL, or any reason other than a pass (ADMIN)

noneffective rate The percentage of a command's members ill or injured as a result of a specific cause, over a given time period (ADMIN)

nonexpendable supplies Materiel that is not consumed through use and, consequently, requires full accountability (LOG)

nonjudicial punishment Administrative disciplinary measures for minor offenses taken by a commander against a soldier who does not demand trial by a court-martial (ADMIN)

nonpay status Period for which a soldier is ineligible to receive pay, due to the individual's own fault or neglect (ADMIN)

nonpersistent agent A chemical that loses its ability to cause casualties within 10 to 15 minutes after its dispersal (NBC)

nonprocurement funds Money budgeted for administrative and maintenance expenditure, not the acquisition of materiel (ADMIN)

nonrecoverable item Materiel consumed through use and not subject to return (LOG)

nonrecurring demand A requisition submitted in support of a one-time requirement; one that does not support demand-based stockage (LOG)

nonselection Determination that an individual is not eligible for promotion or a duty position (ADMIN)

nonstockage list item A supply item authorized for issue by a given activity but not to be maintained in inventory (LOG)

NORAD North American Air Defense Command (AD)

normal bed capacity The planning space for peacetime setup of patient beds in a medical facility, the normal requirement being 100 square feet per bed (MED)

normal interval The lateral spacing between a rank of soldiers equivalent to one extended arm's length (DC)

NPD No pay due; term used on leave and earnings statement to indicate a soldier received no pay (ADMIN)

NR Prosign meaning message number (COMMS)

NSA National Security Agency (INTEL)

NSN National stock number (LOG)

NTC National Training Center (TNG)

nuclear, biological, chemical collection center An agency responsible for collection and dissemination of information on battlefield contamination (NBC)

nuclear warning message (NUCWARN) An alert that must be issued to any friendly forces that may be affected by a planned nuclear detonation (NBC)

nuclear yield The blast energy released by the detonation of a nuclear weapon; measured in the equivalent yield generated by kilotons or megatons of TNT (NBC)

NUCWARN Message warning of a friendly nuclear strike (NBC)

nuisance minefield An area in which mines have been laid to delay/disorganize the enemy (TAC)

null **1.** Character within a message that has no meaning other than to fill a space; a place-holding character (COMMS); **2.** An area or antenna altitude in which no radio signal is received (COMMS)

O

OAC Officer advanced course (TNG)

OADR Originating agency's determination required (INTEL)

oak leaf cluster (OLC) Device signifying the additional award of a service ribbon; worn with the stems pointed to the wearer's lower right (PERS)

O & I Operations and intelligence (ADMIN)

OAS Organization of American States (STRAT)

OB Order of battle (INTEL)

OBC Officer basic course (TNG)

objective The physical goal of an offensive, the seizure of which is essential to the commander's plan (STRAT)

obligated tour A compulsory initial term of service served by other than Regular Army officers (ADMIN)

obligation of funds The legal reserve of money against an appropriation for the purchase of goods or services (ADMIN)

observation post Any position from which there is a substantial view of an area of tactical interest (TAC)

observed fire Points of impact of fire that can be seen and therefore adjusted (ARTY)

observing angle Angle between a line from the target to the gun and a line from the target to an observer (ARTY)

observing interval The time between successive observations of a moving target (ARTY)

observing point The exact position on which fire control personnel sight to obtain firing data (WPN)

obstacle Any natural or artificial barrier that stops, delays, or diverts movement (TAC)

obstacle approach angles The incline formed at the top of a vertically negative obstacle or the base of a vertically positive obstacle that a vehicle must negotiate in overcoming the obstacle (TAC)

OCA Offensive counterair (AVN)

occulter A searchlight's shutter, closed when not in use to prevent targeting by the enemy (WPN)

occupation phase of military government The period that begins when control has been passed to the occupying force capable of enforcing public safety and order (CA)

occupied beds A hospital's patient population as of midnight, including patients on less than a 72-hour pass (MED)

occupied territory An area under the authority of an armed military force rather than civil authorities (STRAT)

OCIE Organizational clothing and individual equipment (LOG)

O club Officers' club (COLL)

OCOKA Observation and fire, concealment and cover, obstacles, key terrain, avenues of approach; criteria for evaluating the military value of terrain (TAC)

OCONUS Outside continental United States (ADMIN)

OCS Officer Candidate School (TNG)

OD Olive drab (green) (COLL)

o dark thirty Early in the morning (COLL)

OEG Operational exposure guidance (NBC)

OER Officer evaluation report (ADMIN)

offensive **1.** A large-scale attack (STRAT); **2.** A ready-to-attack or attacking posture (STRAT)

offensive mine countermeasures Measures that prevent the enemy from laying mines (TAC)

office of record A central point at which records for a specific operation or program are maintained (ADMIN)

officer A soldier holding a commission or warrant from the President or Secretary of the Army conferring the authority to act on behalf of the United States within his or her duty position (PERS)

officers' call Any presentation of information, through briefing or printed material, by a command element to those officers concerned (ADMIN)

off limits Areas or establishments that are not to be entered, by direction of a local commander (ADMIN)

off line A signal processing function performed before or after transmission (COMMS)

offset method A secure means of designating a point on a map by describing the distance from the bottom and then the distance east or west of a north-south line (NAV)

offset plotting Separate determination of range and azimuth data for each gun in a battery to enable it to fire on a moving target (ARTY)

OG Olive green (LOG)

ogive The curved, nose section of a projectile (ORD)

OIC Officer in charge (ADMIN)

OIR Other intelligence requirements (INTEL)

OJT On-the-job training (TNG)

O/L **1.** Observation/losing (ARTY); **2.** Operating level (LOG)

OLC oak leaf cluster (PERS)

old man A unit's commander (COLL)

OM Prosign meaning old man (COMMS)

OMG Operational maneuver group (OPFOR)

omitted cluster A two-meter semicircular area within a mine strip that contains no mines (TAC)

on Command of execution to halt the traverse of a tank's main gun turret; preparatory command is steady (ARM)

one hundred percent rectangle Area in which all rounds from a gun or battery are expected to impact (ARTY)

one-time system A cryptosystem in which the cipher component is used only once, to enhance communications security (COMMS)

on line A signal processing performed during transmission (COMMS)

on-line cryptooperation Equipment that combines the encryption and transmission of messages (COMMS)

OO Immediate message precedence (COMMS)

o/o on order (TAC)

OP **1.** Observation point (TAC); **2.** Prosign meaning operator (COMMS)

OPC Office of primary concern (ADMIN)

OPCON Operational control (ADMIN)

open code A message that appears to be unencrypted plain text (to avoid attempts at decryption) but in fact contains a hidden message (COMMS)

open column A motor march with increased space between vehicles to disperse targets (TAC)

open mess Nonappropriated sundry fund activities that provide essential dining, billeting, and recreational facilities to officers, noncommissioned officers, and their families (LOG)

open ranks Preparatory command (command of execution—march) directing troops in formation to increase the front-to-rear interval between the ranks (DC)

open route A roadway not subject to movement control restrictions (TAC)

open sheaf Lateral distribution of the fire from two or more guns where the adjoining points of impact are separated by the shells' burst radius (ARTY)

open source information Information of potential intelligence value that is available to the public and is, therefore, unclassified (INTEL)

operating level That quantity of materiel necessary for stockage to satisfy demands (LOG)

operating maintenance User-level service and repair of organizational equipment (LOG)

operating program The planning tool under which an organization attempts to match available resources to requirements in support of its annual mission (ADMIN)

operation A military mission, consisting of a specific objective and the maneuver and support required to accomplish it (STRAT)

operational chain of command Temporary lines of authority established for a particular mission or series of maneuvers (ADMIN)

operational command The authority to assign missions, deploy units, and retain or delegate operational control (ADMIN)

operational control (OPCON) The provision of an element to a command for a limited purpose or mission, without also shifting responsibility for administrative, logistical, or other support functions (ADMIN)

operational exposure guidance (OEG) A command's advice to a maneuvering element as to how its mission must be restricted to minimize exposure to radiation (NBC)

operational readiness training High-level collective training designed to maintain units' readiness for their wartime mission (TNG)

operation order (OPORDER) A directive to execute a mission (ADMIN)

operation plan (OPLAN) A projected mission or series of related missions (ADMIN)

operations overlay A map overlay depicting the current deployment of friendly elements (TAC)

operations security (OPSEC) The denial from enemy observation of indicators of friendly intentions and any information related to execution of maneuvers (INTEL)

OPFOR Opposing forces (TNG)

OPLAN Operational plan (TAC)

OPMS Officer personnel management system (ADMIN)

OPORD Operations order (TAC)

opportune lift Cargo space left over after planned transportation requirements have been met (LOG)

opposing force program (OPFOR) Tactical training that focuses on the doctrine of potential adversaries (TNG)

OPSEC Operations security (TAC)

OPTEMPO Operating tempo (TNG)

orange forces The designation for enemy forces during NATO exercises (TNG)

ORB Officer record brief (ADMIN)

order Any communication directing execution of a lawful mission without necessarily specifying the details of execution (MS)

order arms Two-part command directing soldiers to return to the position of attention after rendering a salute (DC)

order for information A directive to subordinate units to provide specified data (INTEL)

orderly room A unit's general business office (COLL)

order of battle The structure of a force; a full description of its personnel strength, command structure, and equipment (INTEL)

order of march A tactical movement plan that includes instructions on speed, security, technique, objective, rally points, sequence of elements, and so forth (ADMIN)

order wire A channel (voice or data) reserved for technical control and maintenance of the system or service (COMMS)

ordinary leave Authorized earned and charged absence from active duty (ADMIN)

ordinary priority The third highest precedence of mission request; below emergency and urgent and above search and attack; used in situations such as when a target delays friendly operations but does not cause casualties (ADMIN)

ordnance Any destructive or explosive materiel such as bombs, rockets, ammunition, and flares (WPN)

Ordnance Corps The combat service support branch and personnel responsible for the development, production, acquisition, and storage of ammunition and weapons (ADMIN)

ORF Operational readiness float (LOG)

organic An element that is an essential part of a unit and is listed in its authorization documents (ADMIN)

organizational maintenance Operator and/or unit-level care for equipment, e.g., inspection, lubrication adjustment, and minor parts replacement (LOG)

organization of the ground The establishment of defensive positions by taking maximum advantage of the terrain and other local conditions (TAC)

Organization of the Joint Chiefs of Staff The command element of the Department of Defense, which includes the office of the chairman, the joint chiefs themselves, and the joint staff (ADMIN)

organization property Materiel that is authorized by tables of organization and equipment or common tables of allowance for assignment to units rather than to individuals (LOG)

organized position A fortified, defensive site (TAC)

orient **1.** To align a map so that its directional arrows point in those actual directions (NAV); **2.** To align a weapon, instrument, or element (NAV)

orientation A familiarization briefing or publication (ADMIN)

originator The individual in whose name or by whose authority a message is sent (COMMS)

origin of the trajectory The center of a gun's muzzle as a projectile leaves it (ARTY)

ORP Objective rally point (TAC)

OST Order ship time (LOG)

OSUT One-station unit training (TNG)

OTRA Other than Regular Army (ADMIN)

out Proword indicating that transmission is complete, no reply expected (COMMS)

outstation (OS) Any station in a communications net other than the control station (COMMS)

OVE On-vehicle equipment (LOG)

over **1.** Proword indicating that a transmission has ended but a reply is expected (COMMS); **2.** Observation of fire indicating excess range—that the round impacted beyond the target (ARTY)

overcast Sky cover of 95 percent or more (AVN)

overpressure Defense of an enclosed space against toxic agents by maintaining it at an air pressure greater than outside atmospheric pressure (NBC)

overprint To fill in standard information on a form to avoid constant rewriting of the same information (ADMIN)

overseas service bars Gold stripes worn near the right cuff on a dress uniform; each bar represents six months of active, OCONUS service (PERS)

overt operation The unconcealed gathering of information (INTEL)

owner-use circuit A communications channel whose use is restricted to a single service (COMMS)

oxidizer The portion of an explosive or propellant that promotes rapid combustion of the explosive or fuel (ORD)

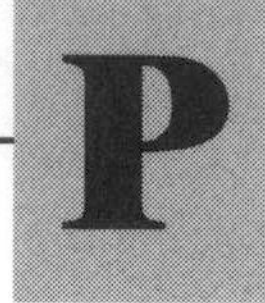

PAC Personnel and Administration Center (ADMIN)

pace count The determination of one's personal number of steps per 100 meters for use in estimating distance traveled (NAV)

pacing items Major weapons or equipment systems of such criticality that they are continuously monitored and intensely managed at all command levels (LOG)

packboard A panel to which gear is secured in order to distribute its weight and make it easier to carry (INF)

packet **1.** A detachment of personnel trained and deployed for a specific mission (ADMIN); **2.** Any assemblage of forms and/or documents submitted to accomplish a standard function (ADMIN)

padding Words or letters added to a message to help conceal its beginning, end, or length (COMMS)

PADS Position and azimuth determining system (ARTY)

PAE Personal arms and equipment (LOG)

panel code Use of highly visible, movable devices for secure, line-of-sight communication (COMMS)

PAO Public affairs office (ADMIN)

parade rest Two-part command directing a position of relaxed attention at which the feet are spread to shoulders' width and hands placed in the small of the back (DC)

parallax error A slight sighting discrepancy that results from the difference between an observing angle and a gun's bore (WPN)

paramilitary Unofficial forces working in parallel or in place of regular armed forces (MS)

parlimentaire An agent of a commander authorized to communicate and/ or negotiate in person with an enemy commander (LAW)

parole A secret control word, to which a sentry responds with the challenge and password (TAC)

partial jurisdiction The sharing of law enforcement authority over territory that is located within both state and federal boundaries (LAW)

partial mobilization Limited activation of Reserve units and individuals to meet the national security requirements resulting from an external threat (STRAT)

partisan A guerrilla fighting covertly against occupying troops (UW)

parts peculiar An item that is restricted to replacement parts from a particular manufacturer (LOG)

PASGT Personnel armor system, ground troop (PERS)

pass Authority for temporary absence from duty not charged against an individual's accrued leave (ADMIN)

passage of lines To attack through a friendly unit that is in contact with the enemy (TAC)

pass in review Combined command directing troops in formation to march past a reviewing officer (DC)

passive Equipment designed to emit no detectable energy and thus avoid enemy surveillance (INTEL)

passive homing Missile guidance based only on emissions from or appearance of the target (WPN)

passive satellite defense The use of cover, concealment, dispersal, camouflage, decoys, and movement to minimize the potential benefit to the enemy from overhead observation (TAC)

pass time The elapsed time between passage of a given point by the first and last vehicles of a column (LOG)

password A secret word that, when used in response to a challenge, establishes the individual's friendly status (TAC)

patching central A facility that interconnects communication channels (COMMS)

pathfinder Individual or team that is airdropped into an area to prepare drop zones and landing zones for follow-on forces (ABN)

path loss Complete attenuation of a radio signal between the transmitting and receiving antennas (COMMS)

PATRIOT Phased array tracking to intercept of target; a very low- to high-altitude surface-to-air missile system (AD)

patrol A small roving detachment whose mission may be reconnaissance, harassment, security, or destruction of an objective (TAC)

pax Passengers (ADMIN)

PBO Property book officer (LOG)

PCO Peacetime contingency operations (STRAT)

PCS **1.** Permanent change of station (ADMIN); **2.** Post, camp, or station (ADMIN)

PD **1.** Priority designator (LOG); **2.** Symbol for the blister agent phenyldichloroarsine (NBC)

P-day The point at which a materiel item is produced at the rate necessary to sustain combat requirements (LOG)

PDO Property disposal officer (LOG)

pecuniary liability Financial responsibility for loss due to fault or neglect (ADMIN)

penetration The breakthrough of a defensive position by an offensive maneuver (TAC)

Pentagon Military headquarters of the United States, located near Washington, D.C. (ADMIN)

percent of slope The rate of rise or fall of the ground over distance; equal to the vertical (elevation) distance multiplied by 100 and divided by the horizontal (ground) distance (TAC)

percussion charge A small, high-explosive charge that is ignited by a firing mechanism and, in turn, ignites a priming charge (ORD)

perfidy Deceptive tactics that tend to destroy the basis for restoration of peace, such as feigning surrender and then attacking (LAW)

perimeter The outer 360-degree edge of a defensive position (TAC)

perimeter defense A defense with flanks protected (TAC)

PERINTREP Periodic intelligence report (INTEL)

PERINTSUM Periodic intelligence summary (INTEL)

periodic intelligence summary (PERINTSUM) A regular intelligence review of a tactical situation produced at corps level or higher, normally at 24-hour intervals (INTEL)

peripheral course Service school training that does not result in award of a military occupational specialty, additional skill identifier, commission, or warrant and is not regarded as career schooling (TNG)

permanent appointment Regular promotion or award of officer's rank in the Regular Army, National Guard, or Reserves, as opposed to temporary, wartime award of such commission (ADMIN)

permanent appropriation A budgetary authorization that remains in effect until revoked by Congress (ADMIN)

permanent change of station (PCS) Reassignment of personnel from one permanent duty location to another (ADMIN)

permanent party Those soldiers assigned to a duty station in continuing support of its mission, as opposed to transient personnel such as trainees (ADMIN)

permanent station A duty post to which a soldier is assigned under orders that do not provide for a termination date (ADMIN)

PERSCOM Personnel command (ADMIN)

Pershing A mobile, long-range, surface-to-surface missile (WPN)

personal locator beacon A radio transmitter designed to provide homing capability in search for individuals separated from friendly forces (NAV)

personal staff Those officers that a commander elects to coordinate directly rather than through his or her chief of staff, typically including aides, the command sergeant major, the inspector general, and the chaplain (ADMIN)

personnel reaction time The time needed by personnel to take protective action after receiving a nuclear strike warning (NBC)

personnel vulnerability condition The degree to which soldiers are exposed to the effects of nuclear weapons (NBC)

PFC Private first class (PERS)

PFR Personal financial record (ADMIN)

PGM Precision guided munitions (WPN)

phase line A line on a battlefield representing the physical progress in control and coordination of the continuing mission (TAC)

phases of military government The evolving degree of control that military forces are able to exercise over occupied territory and its civilians (CA)

phases of smoke The categorization of the generation of obscuring smoke—(a) streamers from the generators, (b) merging of streamers, (c) uniform obscuration, and (d) thinning and loss of effectiveness (TAC)

phases of training The categories of training that individuals and units must progress through to attain operational proficiency—basic combat training, advanced individual training, basic unit training, advanced unit training, field exercises and maneuver training, and operational readiness training (TNG)

phonetic alphabet Distinctive words spoken in place of letters for the purpose of eliminating confusion regarding letters that have a similar sound (COMMS)

phony minefield An area made to appear to contain mines by marking or disturbing the ground, for the purpose of expediently restricting enemy maneuver (TAC)

phosgene (CG) A toxic choking agent; decontaminate by aeration (NBC)

phosgene oxime (CX) A toxic blister agent; decontaminate with water (NBC)

photogrammetry The use of photographs to derive measurements of objects (INTEL)

photographic dosimetry Use of photographic film to measure personal exposure to radiation (NBC)

photographic reconnaissance Air photography designed to provide information on enemy activities and the results of bombings, not for map making (INTEL)

photographic scale The ratio of a measured distance on a reconnaissance photo to the corresponding distance on the ground (INTEL)

phototopography The production of maps depicting surface features from aerial photographs (INTEL)

physical inspection A commander's check made by direct, in-person observation rather than by examination of relevant records or through intermediary personnel (ADMIN)

physical profile serial (PULHES) A numerical summary of a soldier's physical and mental fitness for duty (MED)

physical security Those measures designed to routinely protect personnel and deny unauthorized access to installations, materiel, and information (TAC)

picket fence A PULHES profile of all ones, indicating no physical limitations (MED)

piece A large gun (ARTY)

piecemeal attack Offensive action that develops as units become available to fight it (TAC)

pilferable item Small, desirable item of materiel that is easily stolen (LOG)

pillaring The undesired, rapid ascent of smoke meant to obscure ground operations (TAC)

pillbox A small, heavily fortified defensive positioning, typically containing heavy machine guns or antitank weapons (TAC)

pinpoint distribution The system for issuance of publications that attempts to match issues and updates with a unit's mission and equipment (ADMIN)

pinpoint target A target less than 50 meters in diameter (ARTY)

pintle Vertical bearing upon which a gun carriage rotates (ARTY)

PIR Priority Intelligence Requirement (INTEL)

PL Phase line (TAC)

plaindress A message in which the originator and addressee are indicated separately from the text (COMMS)

plain text An unencrypted message or signal (COMMS)

planning factor A multiplier used to estimate an operation's combat service support requirements (LOG)

planograph A scale diagram of a storage layout, showing aisles, walls, doorways, bin areas, rack areas, pallet areas, and other support areas in detail (LOG)

plan position indicator The display screen that indicates the relative position of objects detected by radar (INTEL)

plastic explosive Charge material that can be easily shaped at normal temperatures (ORD)

plasticizer A material added to a high explosive to make it less brittle (ORD)

platform rock The backward and upward movement of a tank upon firing of the main gun, which causes dislocation of the sights (ARM)

platoon An administrative subdivision of a company, battery, or troop; further subdivided into squads (ADMIN)

PLDC Primary Leadership Development Course (TNG)

PLF Parachute landing fall (ABN)

PLL Prescribed load list (LOG)

plotting and relocating board A scale model of a battery's field of fire; used for planning, computation, and control of fire (ARTY)

plunging fire Rounds that impact the ground at a high angle (WPN)

PMCS Preventative Maintenance Checks and Services (LOG)

PMOS Primary military occupational specialty (ADMIN)

POC Point of contact (ADMIN)

POE Port of embarkation (LOG)

pogie bait Snack food (COLL)

POI Program(s) of instruction (TNG)

point designation grid A system of arbitrary reference lines overlaid on a map, chart, or photograph, designed to provide a secure means of navigational planning (NAV)

point feature An object, the location of which can be described by a single set of coordinates (TAC)

point man Soldier at the front center of a tactical movement; his purpose is to provide advance security and reconnaissance (TAC)

point target A small but critical target requiring exact placement of fire for its destruction (ARTY)

POL Petroleum, oils, lubricants (LOG)

polar coordinates A target position derived from a known observer's position and his or her determination of range and azimuth (ARTY)

police To clean or straighten up; to attend to unfinished details (COLL)

POMCUS Positioning of material configured to unit sets (LOG)

pool equipment Materiel that, because of its intermittent need, is held in reserve for issue to units as their missions require (LOG)

port arms Two-part command directing a soldier to hold his or her rifle barrel up, diagonally, on a line from the left shoulder to the right hip (DC)

port call Request from a transporting agency for movement of personnel or cargo to the loading area (LOG)

position area survey The determination of the exact horizontal and vertical location and orientation of a battery (ARTY)

position correction Adjustment of firing data to compensate for the offset distance between a gun and the base piece of its battery (ARTY)

position defense Strategy that relies on the ability of well-placed defenders to control the terrain and, consequently, the battle (TAC)

positioning band Metal band on some recoilless ammunition that assists in the seating of the round in the chamber (ORD)

POSNO Position number (ADMIN)

posse comitatus act Laws that generally prohibit the military from operating in support of civilian law enforcement (LAW)

post The positions taken by the officer in charge and NCO in charge in front of a formation (DC)

postamble That section of a message immediately following the text (COMMS)

post exchange (PX) Supply source for personal, civilian items (LOG)

post flag A nylon national flag measuring 10 feet by 19 feet that is displayed under general circumstances in fair weather (DC)

POV Privately owned vehicle (ADMIN)

PP Priority message precedence (COMMS)

pralidoxime chloride A chemical used to counteract the effects of a nerve agent (MED)

prearranged fire Scheduled fire on targets of known location, normally planned well in advance (TAC)

preassault operation Actions taken against the enemy in preparation for offensive operations, such as reconnaissance, bombardment, and mine clearing (TAC)

precedence A designation of the handling priority a message is to receive relative to other communications traffic (COMMS)

precision fire Fire direction in which the rounds' center of impact is on a point or limited target (ARTY)

preclusive buying Purchase of materiel to prevent its acquisition by the enemy (LOG)

precursor A shock wave that precedes the main blast wave resulting from a nuclear detonation; caused by air displacement from the main blast (NBC)

predicted firing Fire delivered on the future calculated position of a moving target, without benefit of adjustment based on observation (ARTY)

predicted point Calculated position of a moving target at the instant of firing, based on its last observation (ARTY)

predicted position That point at which a moving target is expected to arrive by the time the projectile reaches it (ARTY)

preemptive attack An offensive action launched because of evidence that the enemy is about to attack (TAC)

preparation fire Artillery or other bombardment of an objective area just prior to assault; intended to destroy and disrupt the enemy defense; also called prep or preparation (TAC)

preparatory command The first half of a combined command; given to alert troops as to what to do upon being given the command of execution (DC)

prepared missile A missile ready to fire, lacking only targeting input (WPN)

preposition To emplace units and/or equipment close to where they will be needed in order to reduce reaction time (TAC)

prescribed load list (PLL) Combat-essential repair parts and supplies authorized to be held at unit level in support of assigned equipment (LOG)

present arms Two-part command directing soldiers in formation to render a salute (DC)

preset guidance Missile navigation control that is set prior to launch and cannot be changed in flight (AD)

pressure mine An emplaced charge designed to detonate on application of weight or pressure (ORD)

preventive maintenance (PM) Routine service of equipment to extend its useful life or to avoid having to maintain it during operations (LOG)

Primacord Detonating cord; a brand name for an explosive cord designed to transmit shock to remote charges (ORD)

primary censorship Military editing, for reasons of operational security, of soldiers' personal communications (ADMIN)

primary radar A target-detection and ranging system based on the transmission of a signal and the sensing of the reflection of that same signal (INTEL)

primary road An artificial surface suitable for vehicular traffic and more than six meters wide (LOG)

primary zone Promotion status of those soldiers fully qualified for promotion (ADMIN)

primer A device that initiates an explosive train (ORD)

priming charge An explosive that transmits the detonation wave to the main charge (ORD)

priority designator A two-digit code resulting from the system for determining the sequence in which units are issued materiel, based on their need and mission (LOG)

priority intelligence requirements Those critical intelligence objectives that a commander chooses to emphasize given limited time and collection assets (INTEL)

priority message (PP) Handling precedence above routine and below immediate (COMMS)

priority of immediate mission A system for determining the relative criticality and, therefore, order of execution of support missions (ADMIN)

prisoner of war (POW) A combatant, or an individual accompanying a combatant force, who is detained by an opposing force and is entitled to treatment specified by the Geneva Conventions (LAW)

private (PVT) An enlisted soldier in the pay grade of either E-1 or E-2 (PERS)

private first class (PFC) An enlisted soldier in the pay grade E-3 (PERS)

proactive To initiate action when its need is apparent rather than required (COLL)

procedure word (proword) Verbal shorthand or code word that permits radio and telephonic conversations to be conducted briefly and with heightened security; also known as procedure sign (COMMS)

production base temperature Classification describing the current operating level on D-day of a materiel-producing facility compared to its capacity; a "hot" facility is producing at its maximum rate, a "cold" facility is available but not producing, and a "warm" one is somewhere above a minimum sustaining rate (LOG)

proficiency flying status The authorization, under competent order, for aviators to maintain flight skills through continuing practice, even though current duties do not require flying (AVN)

proficiency pay Additional pay authorized to soldiers who maintain special perishable skills (ADMIN)

proficiency rating designator A code that follows an enlisted rank and specifies the proficiency pay to which a soldier is entitled (ADMIN)

profile Certification of a physical limitation for medical reasons (ADMIN)

programmed strength A planned level of manning and equipment fill to attain a prescribed degree of readiness for future operations (ADMIN)

projectile Any object that is propelled through the air by an external force and continues by its resulting inertia for a destructive purpose (WPN)

promotable Status designation of an individual who has been selected for promotion but whose position on the eligibility list has not yet been reached (PERS)

promotion list A roster of those officers or noncommissioned officers selected for promotion, in the order of their standing (ADMIN)

prone position An individual firing position; flat on the ground, chest down (TAC)

proofing The testing of a suspected mined area to ensure that any mines present have been neutralized (TAC)

propaganda Any information intended to influence the opinions or emotions of a targeted audience in support of national or military objectives (UW)

propagation mode Type of path taken by radio waves from transmitter to receiver (COMMS)

propelling charge A charge that produces explosive gases to fire a projectile (ORD)

property book A compilation of the inventory and information required to manage the individual items belonging to a unit (LOG)

property book accountability The obligation of a command to maintain inventory records and control of its nonexpendable materiel (LOG)

property disposal An installation activity through which damaged or obsolete materiel is redistributed among units or sold; a salvage activity (LOG)

proponent Any staff, school, or organization assigned the responsibility for development of a specific area of doctrine or training (ADMIN)

prosign Procedure sign; *See* procedure word (COMMS)

protected frequency A portion of the radio communications spectrum that is not to be jammed because friendly forces are monitoring enemy use of it (EW)

protected persons Those civilian nationals who fall into the hands of the force opposing their nation, and whose treatment is governed by the Geneva Conventions (LAW)

protective mask Device that covers an individual's face and filters inhaled air of chemical and biological agents and radioactive particles (NBC)

protective minefield A temporary placement of mines in support of a defensive position (TAC)

provisional unit A command assembled for a specific, limited mission (ADMIN)

provost court A military tribunal with limited jurisdiction over an occupied territory (LAW)

provost marshal Staff officer responsible for a command's military police and prisoner-of-war activity (ADMIN)

proword Procedure word (COMMS)

proximity fuse A fuse that initiates detonation based on closeness of the round to the target (ORD)

PSSP Personnel Security and Surety Program (INTEL)

psychological consolidation Psychological operations in support of military objectives directed at the civilian populace under friendly control (UW)

psychological operations (PSYOPS) The full range of political, military, economic, and ideological activities designed to affect the emotions, attitudes, or behavior of friendly or enemy audiences in support of national objectives (UW)

psychological warfare The use of propaganda to negatively influence the enemy's willingness to fight (UW)

PSYOPs Psychological operations (UW)

PSYWAR Psychological warfare (UW)

PT Physical training (TNG)

PTL Primary target lives (ARTY)

public affairs The activity responsible for community relations and dissemination of internal information to the public (ADMIN)

PUC Presidential Unit Citation (ADMIN)

pugil sticks Padded clubs used in training aggressiveness and hand-to-hand combat (TNG)

PULHES Physical capacity, upper extremities, lower extremities, hearing, eyes, neuropsychiatric; physical profile estimates (MED)

purple forces The designation for forces in opposition to both blue (friendly) and orange (enemy) forces during NATO exercises (TNG)

Purple Heart Decoration in recognition of wounds received in action against an enemy (PERS)

pursuit An offensive action meant to overtake and destroy a retreating enemy (TAC)

PVO The Soviet air force (OPFOR)

PX Post exchange (LOG)

pyrotechnics Signaling devices that detonate to produce light, heat, or sound (ORD)

PZ Pickup zone (TAC)

Q

Q and Z signals A series of brevity codes used primarily on continuous wave (Morse) channels; *See* Table A-3 (COMMS)

QD Quantity distance (LOG)

QM **1.** Quartermaster (LOG); **2.** Quartermaster Corps (LOG)

QMP Qualitative management program (ADMIN)

QSS Quick supply store (LOG)

QSTAG Quadripartite Standardization Agreement (ADMIN)

quad 50 A fighting vehicle equipped with four integrated .50 caliber machine guns (WPN)

quantity/distance tables Regulations specifying the standards for storage of explosive munitions (ORD)

Quartermaster Corps (QM) The branch and soldiers responsible for supply of materiel (LOG)

quarters Lodging or housing (LOG)

quarters in kind Lodging provided without cost (LOG)

quick fire Small arms fire delivered immediately on a fleeting target (TAC)

Quickfix The AN/ALQ-151 rotary-wing intercept, direction-finding, and jamming aircraft (EW)

Quicklook A fixed-wing electronics intelligence collection and radio-direction-finding aircraft (EW)

quickmatch A rapidly burning fuse, impregnated with black powder (ORD)

quick release Rigging of equipment that permits its immediate jettison (ABN)

quick smoke An unplanned mission to build a small smoke screen of short duration (ARTY)

quick time Marching at a rate of 120 paces per minute (DC)

quota A reserved slot in a training program (TNG)

RA Regular Army (ADMIN)

RAAMS Remote antiarmor mine system (ORD)

racer A totable gun carriage (ARTY)

RACO Rear area combat operations (TAC)

rad Obsolete term for the unit of absorbed radioactivity; *See* centigray (NBC)

radar Radio detection and ranging; a target location system (INTEL)

radar picket An outlying radar site away from the force to be protected (INTEL)

radar ranging Use of radar to determine distance to target (INTEL)

radar silence The temporary prohibition of radar emissions for the purpose of denying the enemy intelligence they could derive from the operation (TAC)

radiac Radioactivity detection indication and computation (NBC)

radiac meter A device for measuring radioactivity (NBC)

radial A discrete magnetic bearing, transmitted from a very high frequency beacon (NAV)

radiation exposure state History of a unit or an individual's previous exposure to radioactivity: Categorized as follows: RES-0—no previous exposure; RES-1—exposure to more than zero centigrays but less than or equal to 70; RES-2—exposure to more than 70 centigrays but less than 150; RES-3—exposure to greater than 150 centigrays (NBC)

RADINT Radar intelligence (INTEL)

radioactivity The emission of subatomic particles, normally resulting from the use of nuclear weapons (NBC)

radio direction finding The determination of a transmitter's direction relative to a receiver (INTEL)

radio guard A receiving station's mission to monitor certain frequencies or types of transmissions (EW)

radio location Determination of a transmitter's position based on analysis of its broadcasts (EW)

radiological Radioactive; concerning emission of nuclear radiation (NBC)

radiological agent A substance that produces casualties by emitting nuclear radiation (NBC)

radio procedure Standard techniques used to promote efficiency of transmitted communications (COMMS)

radio relay system High frequency, line-of-sight links used to transmit voice and teletype channels beyond their normal range without the use of trunk wire circuits (COMMS)

radio silence The order to not transmit for a specified time period (COMMS)

radius of action The distance that a combat loaded vehicle or aircraft can travel, perform its mission, and return without refueling (TAC)

radius of damage The area around a projectile's impact within which there is a 50 percent probability of achieving the desired level of destruction (NBC)

radome The protective cover around a radar antenna (INTEL)

RAG Regimental artillery group (OPFOR)

raging bull message An alert notification that is not a test (ADMIN)

raid Small-scale offensive operation with a specific, limited objective other than to secure ground; followed by planned withdrawal (TAC)

raid clerk The soldier in an aircraft warning service responsible for initial identification of flights (AD)

railhead A point for cargo transfer between a railway line and other transport means (LOG)

rally point A reassembly point planned for members of a tactical unit who may become separated during an operation (TAC)

random mixed alphabet Cipher key in which any letter is as likely as another to be next in the sequence (COMMS)

range **1.** The distance to a target (TAC); **2.** An area prepared for practicing live weapons fire or maneuver (TNG)

R & R Rest and recreation; a break from duty status (COLL)

range card Any chart showing distances to targets on a live-fire training area or, for heavy guns, the amount of charge required to reach the targets (TNG)

range flag Red banner flown as a warning that a live-fire training exercise is being conducted (TNG)

range K A correction for nonstandard conditions, expressed in meters per 1,000 meters of range (ARTY)

range quadrant Instrument used to measure a gun's elevation (ARTY)

Ranger A soldier specially trained to conduct small raids, irregular operations, and long-range reconnaissance (PERS)

range table The portion of a firing table that correlates elevation and range under various conditions (ARTY)

ranging The process of determining the distance to a target (ARTY)

rank **1.** A soldier's relative level in the chain of command and personnel system (ADMIN); **2.** A line that is one element deep (DC)

RAP **1.** Rear area protection (TAC); **2.** Rocket-assisted projectile (ARTY)

rappel To descend in a controlled slide down a rope, either from a hovering helicopter or down a sheer plane such as a cliff or a building (TAC)

RAS Regimental aviation squadron (ADMIN)

ratelo Radiotelephone operator (COMMS)

rate of fire The number of rounds a weapon is theoretically capable of firing per minute, disregarding overheating and maintenance (WPN)

rate of march An element's average speed of nontactical movement, including halts (TAC)

rater The first-level evaluator in an efficiency assessment system (ADMIN)

ration Food for one soldier for one day (LOG)

rationalization The reallocation and standardization of defense resources to improve the effectiveness of allied forces (ADMIN)

ration cycle A day's worth of (three) meals, beginning with any of them (LOG)

ratline An organized effort for clandestine cross-border movement of people or materiel (TAC)

RATT Radio teletypewriter (COMMS)

raw score The number of correct responses given during a test (TNG)

RC Reserve component (ADMIN)

RDD Required delivery date (LOG)

RDF Radio direction finder (TAC)

RDX Cyclonite, a high explosive (ORD)

readiness The current ability of forces or systems to deploy and perform their planned mission (ADMIN)

reading file File of background information on a command's current situation maintained for reference by whoever needs to understand the unit's status (ADMIN)

read on/off The briefing/debriefing of intelligence personnel regarding precautionary measures to be taken in safeguarding classified information (INTEL)

ready Status of a weapon that is loaded, aimed, and prepared to fire (WPN)

ready missile A guided tactical projectile that is on its launcher, awaiting the order to fire (WPN)

Ready Reserve Mobilization Reinforcement Pool Soldiers assigned to Control Group (annual training), Control Group (reinforcement), or the faculty and staff of U.S. Army Reserve Schools (ADMIN)

reallocation authority The authority of NATO commanders to redistribute national logistic resources under their command during wartime (LOG)

real time The absence of delay between the occurrence of an event and the electromagnetic transmission of data concerning it (INTEL)

rear area The sector away from the enemy, behind combat and forward areas, where combat service support elements are located (TAC)

rear echelon Elements of a force that are not required to secure and hold an objective area (TAC)

rear guard The element that provides security, through reconnaissance, delay, and harassment, in the direction opposite a unit's movement (TAC)

rear party The portion of the rear guard that protects the withdrawal of the rear point (TAC)

rear point A moving unit's rearmost trailing echelon, whose purpose is to observe and discourage enemy pursuit (TAC)

REC Radio-electronic combat (OPFOR)

recapitulation Consolidated listing of a unit's personnel and materiel assets (ADMIN)

recce Reconnaissance (COLL)

receiver The section of a weapon that accepts a round from the magazine prior to its seating in the breech (WPN)

reception station A facility for the inprocessing of new recruits (ADMIN)

rechamber The reboring of a weapon's chamber (WPN)

reciprocal laying Method of setting two guns so that their planes of fire are parallel (ARTY)

recoil The rearward motion of a weapon that results from the firing of an explosive projectile (WPN)

recoilless weapon A weapon in which the rearward force of firing is dampened through the use of gas ports (WPN)

recoil mechanism Any device that absorbs the rearward shock when a round is fired (WPN)

recoil-operated weapon A weapon that ejects a spent round, loads the next round, and positions the firing mechanisms from the force of the gas pressure generated by the firing of the first round (WPN)

reconnaissance The observation of an area or item of military significance, especially troop and equipment deployment, terrain, and weather (INTEL)

reconnaissance by fire Fire directed at suspected enemy positions for the purpose of forcing them out into sight (TAC)

reconnaissance in force A detachment whose primary mission is observation of the enemy and terrain but that is also deployed in sufficient strength to engage likely targets of opportunity (TAC)

reconsignment The redirection of cargo and/or materiel to a new destination (LOG)

reconstitution site A site at which a headquarters is to be re-formed after destruction of the original headquarters (ADMIN)

record firing A grading of a soldier's marksmanship ability (ADMIN)

recover **1.** To resume one's original position or the position of attention (DC); **2.** To reconstruct plain text from encrypted text (INTEL)

recovery The extrication of a vehicle from any position in which it is unable to proceed (LOG)

recovery vehicle The M88A1, a fully tracked vehicle designed to move a vehicle that is stuck or disabled (LOG)

rectangle of dispersion An area within which the rounds fired from a gun will fall under identical firing conditions (ARTY)

recuperability The ability of a damaged target to return to operation given the limitations of time, equipment, and personnel (AD)

recurring demand A requirement for materiel that is regular and can therefore be planned for (LOG)

red concept Circuitry carrying classified, plain-language message traffic (COMMS)

redeploy To change locations; to return to one's permanent duty station (COLL)

Redeye The FIM43A, a short-range shoulder-fired, man-portable, surface-to-air missile (WPN)

Redleg An artilleryman (COLL)

redoubt A fortress or stronghold (TAC)

REDTRAIN Readiness training (ADMIN)

reduction coefficient The ratio of observer/target range to gun/target range, which is multiplied by the observed deviation to obtain corrected firing data (ARTY)

refer To aim an artillery piece without changing its lay (ARTY)

reference piece The gun that serves as the standard for comparison among those guns in a battery (ARTY)

Reforger Return of forces to Germany (TNG)

refugee A civilian who flees conflict in his or her home area for sanctuary elsewhere (ADMIN)

regiment A unit at a command level below a division and above a battalion (ADMIN)

regional chart A 1:1,000,000 scale map used in air navigation (AVN)

registered document A numbered, short-titled, classified document for which accountability is maintained (ADMIN)

register guide Compilations of graphics, music, sound effects, and other symbols thought to be psychologically effective against a given target population (UW)

registration fire Fire delivered for the purpose of deriving data for later targeting (ARTY)

Regular Army (RA) The permanent—peace and wartime—component of the U.S. Army (ADMIN)

regulating station A logistical movement control agency (ADMIN)

reinforcement training unit A Reserve unit that provides nonpay status training for members attached from the Ready Reserve Mobilization Reinforcement Pool (TNG)

relateral tell The relay of information through a third party when direct communication is not possible (INTEL)

relative effectiveness (RE) A measure of the shattering power of an explosive, as compared to TNT (ORD)

release point **1.** A point along a road movement at which the elements of a column separate and proceed under control of their respective commanders (TAC); **2.** The location at which personnel or materiel are airdropped (ABN)

release unit An item of materiel that, because of transportability considerations, must be offered to a movement control authority for shipment (LOG)

releasing officer Soldier authorized to issue a message in the name of an originator (COMMS)

relief **1.** Release from a duty position (ADMIN); **2.** The shape and height of the land forming the earth's surface (NAV)

relief in place Continuation of a mission by an element that assumes the position of the element being relieved (ADMIN)

relocation Determination of range and azimuth to a target based on either the range/azimuth from another station or, in the case of a moving target, the previous position (ARTY)

REMBASS Remotely monitored battlefield sensor system (INTEL)

REMS Remotely employed sensor (INTEL)

repair-cycle float Mission-essential equipment authorized in excess of operational needs to substitute for like units during scheduled repair (LOG)

repatriation The return to one's native country after having left it under duress or threat of conflict (LAW)

replacement company A holding unit that administers and trains soldiers prior to their assignment to units (ADMIN)

replacement factor An estimate of the percentage of equipment that will require replacement as a result of loss, theft, wearout, or enemy action (LOG)

replacement stream input Soldiers who have entered the Army, but are in processing and training and not assigned to permanent duty (ADMIN)

report line Point at which moving troops must communicate with their command echelon (TAC)

report of survey An official description of the circumstances surrounding loss or destruction of government property that serves as authority to drop the property from accountability and affixes responsibility and liability for charges (LOG)

request for issue or turn-in DA Form 3161, a multipurpose form used to manage the flow of materiel in the supply system (LOG)

required strength The full or level-one manning level under a modification table of organization and equipment (ADMIN)

requirements control Regulation of the initiation and preparation of reports (ADMIN)

resection Determination of one's position by plotting on a map the intersection of azimuths derived from compass bearings to two or more known locations (NAV)

reserve **1.** Combat forces held to the rear of the battle area, uncommitted, until their deployment would be decisive to the outcome (TAC); **2.** Soldiers not on active duty but subject to activation (ADMIN)

Reserve Components The Army National Guard of the United States, the Army Reserve, the Naval Reserve, the Marine Corps Reserve, the Air National Guard of the United States, the Air Force Reserve, and the Coast Guard Reserve (ADMIN)

reserved area A sector designated as significant to the national defense by the President, access to which is restricted (TAC)

reserved demolition target A target critical to an operational plan, the destruction of which is closely controlled in synchronization with that plan (TAC)

Reserve Officers' Training Corps (ROTC) Organization within the civilian education system that provides military training to prospective officers (TNG)

residual contamination Nuclear, chemical, or biological agents that remain even after a decontamination procedure; this residue is normally acceptable (NBC)

residual radiation Radiation that persists longer than one minute after a nuclear burst (NBC)

resolution The ability of imaging equipment to distinguish an object from its background (INTEL)

resolution in azimuth The angle by which targets at the same range must be separated in order to be distinguishable by a given radar set (INTEL)

resolution in range The distance by which targets on the same azimuth must be separated in range to be distinguishable by a given radar set (INTEL)

responsible officer A soldier assigned by law or regulation to discharge a duty; may be a commissioned officer or an NCO (ADMIN)

rest Combined command directing troops in formation to keep their right foot in place but permitting movement and conversation (DC)

restitution The transfer of target location information from aerial photographs to maps (INTEL)

restricted data Information concerning the design and manufacture of nuclear weapons or the production of nuclear material (WPN)

restricted frequency A portion of the radio communications spectrum that may be jammed only after coordination with friendly users (EW)

restriction Nonjudicial punishment that sets limits on a soldier's movement (LAWS)

restrictive fire area/restrictive fire line A sector or point beyond which fire or certain types of fire cannot be delivered without coordination with the establishing command (TAC)

resubordination The transfer of units from the authority of one command to another based on changing operational needs (ADMIN)

retained enemy personnel Humanitarian assistance personnel (e.g., medical personnel and chaplains) held with prisoners but not as prisoners (ADMIN)

reticle The sighting reference marks inside a viewing apparatus (WPN)

retirement **1.** Movement away from the battle area by a force that is already out of contact (TAC); **2.** Leaving the Army after 20 or more years of service with a guaranteed pension for life (ADMIN)

retransmission, retrans The simultaneous reception, and transmission on a different frequency, of a radio signal that would otherwise be beyond the range of elements that need to maintain contact (COMMS)

retreat The evening flag-lowering ceremony and the accompanying bugle call (DC)

retrofit Modification of in-service equipment to enhance or update its performance (LOG)

retrograde Movement to the rear, away from the enemy; any movement opposite the normal flow (TAC)

reutilization assignment Reassignment of a soldier with specialized training to a second duty position requiring that training (ADMIN)

reveille The morning flag-raising ceremony and accompanying bugle call (DC)

reversed standard alphabet Cipher key that is in backward alphabetical order (COMMS)

reverse slope Terrain that descends away from the enemy (TAC)

revetment A facing of stone, cement, or sandbags designed to protect against the effects of shrapnel or strafing (TAC)

review A ceremonial formation, normally of a battalion or larger (DC)

reviewing authority **1.** Officer or agency required to affirm the findings of a court-martial before the sentence can be carried out (LAW); **2.** Office authorized to execute final action on reports of survey (LOG)

RF **1.** Radio frequency (COMMS); **2.** Representative fraction (NAV)

RFL Restrictive fire line (TAC)

RFO Request for orders (ADMIN)

RHA Records holding area (ADMIN)

ribbon bridge A floating assault bridge (ENG)

RICC Reportable item control code (LOG)

RIF Reduction in force (ADMIN)

right face A two-part command (preparatory: right; execution: face) directing troops in formation to execute a 90-degree turn to the right (DC)

riot gun A short-barrelled shotgun suitable for crowd dispersal (WPN)

rip stop Fabric that is reinforced with nylon cord to prevent tears from spreading (LOG)

risers Connecting straps between the personal parachute harness and suspension lines to the canopy (ABN)

riverine area An area with extensive water surface and/or inland waterways suitable for transportation and limited land lines of communication (TAC)

R method Type of message transmission that requires the receiver to acknowledge receipt (COMMS)

RO Requisitioning objective (LOG)

road clearance time The time required for a column to pass a given point (LOG)

road guard A soldier detailed to control vehicular traffic around marching formations (ADMIN)

road wheel A wheel of an armored vehicle that is in contact with the ground through the track (ARM)

rocket A self-propelled projectile containing its own fuel, and oxidizer if necessary, that is not guided after launch (WPN)

rocket launcher Any platform or device that provides stability during a rocket's ignition phase (WPN)

roger Oral acknowledgment that a transmission has been received and was understood (COMMS)

ROK Republic of Korea (ADMIN)

ROLAND Short-range, low-altitude, all-weather air defense missile system (AD)

roll up Reassembly of a squad- or platoon-size element that has been dispersed, as in an airborne operation (TAC)

RON Remain overnight (ADMIN)

room circuit A self-contained encipherment/decipherment circuit that is not connected to outstations (COMMS)

ROP Reorder point (LOG)

ROPS Roll-over protection system (LOG)

rossette A lapel device worn on civilian clothing consisting of the gathered suspension ribbon of a medal (PERS)

rotary wing A helicopter-type aircraft (AVN)

ROTC Reserve Officers' Training Corps (ADMIN)

round A complete unit of ammunition necessary to fire a weapon once, consisting of primer, propellant, and projectile (ORD)

Round House Exercise term for DEFCON 3, a state of alert (STRAT)

roundout The employment of Reserve component units to fill understructured active divisions to NATO standards (ADMIN)

rounds complete Report that a specified number of rounds have been fired (ARTY)

route reconnaissance Mission to gather information about and/or along a specified line of communication (INTEL)

route step Preparatory command for troop marching when their only requirement is to attempt to maintain formation permissible for long-distance or rough-terrain marches (DC)

route transposition Encryption using a specified pattern through a cipher matrix to accomplish the substitution (COMMS)

routine message (RR) Lowest level precedence, (below priority) indicating ordinary handling (COMMS)

routing indicator An alphabetic code that directs communications or materiel to a specific command or fixed location (ADMIN)

roving field artillery Guns that are moved around the battlefield for the specific purpose of deceiving the enemy as to position and strength of friendly forces (TAC)

RPG Rocket-propelled grenade (WPN)

RPV Remotely piloted vehicle (AVN)

RR Routine message precedence (COMMS)

RSR **1.** Resource status report (LOG); **2.** Required supply rate (LOG)

RST Regularly scheduled training (ADMIN)

rucksack Backpack for personal equipment (INF)

rules of engagement Directives specifying the circumstances and limitations under which military forces will initiate and/or maintain combat with the enemy (MS)

running spare Repair part that is shipped along with a major system or end item because of its frequent usage (MAINT)

RYE Retirement year ending (ADMIN)

RZ Recovery zone (TAC)

S

S1 Adjutant, battalion level and below (ADMIN)

S2 Intelligence officer, battalion level and below (ADMIN)

S3 Training and/or operations officer, battalion level and below (ADMIN)

S4 Supply officer, battalion level and below (ADMIN)

S5 Civil affairs officer, battalion level and below (ADMIN)

SA Symbol for the blood agent arsine (NBC)

sabot Jacket around a projectile that permits its firing through a larger-caliber bore (ORD)

sabotage Intentional clandestine destruction of assets by enemy agents (UW)

saddle The lower ground between two hilltops or a break along an otherwise level ridge crest (NAV)

SAEDA Subversion and espionage directed against the U.S. Army and deliberate security violations (INTEL)

safe Set so as not to detonate; disabled or unarmed (WPN)

safe conduct Document issued by a military authority that affords access to a restricted area (ADMIN)

safeguard A written order for protection of specified enemy persons or property (ADMIN)

safe house A small base for clandestine operations, the primary security for which is its appearance as a civilian residence (INTEL)

safety diagram A graphic representation of those areas into which fire can be delivered (TAC)

sagger Soviet-bloc wire-guided antitank weapon with a shaped-charge HEAT warhead (WPN)

SALUTE report A reconnaissance/intelligence report of visual contact with an enemy element in the following format: size, activity, location, unit, time, and equipment (INTEL)

saluting distance The distance—up to 30 paces—at which rank is readily visible and a hand salute is to be rendered to a superior officer (ADMIN)

salvo The simultaneous fire of a number of weapons against a target (ARTY)

SAM Surface-to-air missile (AD)

sand table Three-dimensional model of a battle or training area; used in operational planning (TNG)

sanitize To remove classified portions of a document (INTEL)

sapper A soldier who lays mines or places charges (PERS)

sarin (GB) Highly toxic nerve agent; decontaminate with STB slurry, bleach, DS 2, hot soapy water, or individual decontamination kit (NBC)

satchel charge A bundle of blocks of explosive material fastened together and fitted with a carrying handle (ORD)

SATO Scheduled airline ticket office (ADMIN)

SAW Squad automatic weapon (WPN)

SB Supply bulletin (LOG)

SBI Special background investigation (INTEL)

SBP Survivor Benefit Plan (ADMIN)

scale factor A correction to a distance measured on a map; necessary because of the slight distortion inherent in all maps (NAV)

SCATMINWARN Scatterable minefield warning (ORD)

scheduled target An enemy location to be fired on at a particular time (ARTY)

scheme of maneuver The tactical plan for the movement required for a force to seize an objective (TAC)

SCI Special compartmented intelligence (INTEL)

SCIF Sensitive compartmented information facility (INTEL)

S-day The date on which the first manpower mobilization action for an operation occurs, if different from M-day; a planning point within the wartime manpower planning system (ADMIN)

SDD Standard delivery date (LOG)

SEAD Suppression of enemy air defenses (TAC)

search and attack priority Lowest precedence of mission request; below ordinary—reserved for fleeting targets that have destructive potential (ADMIN)

searching fire The distribution in depth of fire by making successive changes in a gun's elevation (TAC)

SECDEF The Secretary of Defense (ADMIN)

second 1/3,600 of a degree of angular measure; also, 1/60 of a minute of angular measure (NAV)

secondary censorship Military editing, for security purposes, of the personal communications of support personnel accompanying regular military units (ADMIN)

secondary item Materiel other than major end items, such as consumable supplies and replacement parts (LOG)

secondary radar A target-detection and ranging system that transmits an interrogator signal that activates a response signal broadcast back from friendly craft (INTEL)

secondary station **1.** A link in a network other than the net control station (COMMS); **2.** An observation post at the end of a gun's farthest baseline (ARTY)

secondary zone The status of those soldiers qualified for promotion but lacking in time in grade or service (ADMIN)

second lieutenant An officer; pay grade O-1 (PERS)

second strike capability The ability to survive an enemy offensive with enough resources intact to launch a counteroffensive (STRAT)

secret A classification for national security material requiring a substantial degree of protection, the unauthorized disclosure of which could cause serious damage to national security (INTEL)

Secretary of the Army The executive head of the Department of the Army, reporting directly to the Secretary of Defense (ADMIN)

section A unit smaller than a platoon and larger than a squad (ADMIN)

sector A unit's designated area of responsibility (ADMIN)

sector of fire An enemy area to be covered by the friendly fire from a particular weapon or unit (TAC)

secure **1.** To gain possession and control of an objective, denying its use or destruction by the enemy (TAC); **2.** To prevent access to classified information by unauthorized personnel (INTEL)

security The protection of any asset against compromise, damage, or destruction (STRAT)

security assistance Programs under which the United States provides materiel, training, and other services to foreign governments in support of mutual policy objectives (STRAT)

security certification Documentation that an investigation of an individual has been made and no disqualifying information found (INTEL)

security classification The categorization of defense information according to the degree of damage its disclosure would cause to national security (INTEL)

security clearance The determination that an individual is eligible for access to certain classified information (INTEL)

SED Simulative electronic deception (EW)

sedition Disloyalty or incitement against a government in power (LAW)

Selected Reserve Ready Reserve active units, individual control group soldiers obligated for periodic training, and Reservists performing initial active duty for training (ADMIN)

selective mobilization Limited activation of Reserve units and individuals to support a domestic emergency not resulting from enemy attack (STRAT)

self-authentication A transmitting station's verification of its identity without interaction with other stations (COMMS)

self-help branch Support activity that supplies construction or remodeling supplies on the basis of submission of an approved plan for their use (ADMIN)

self-service supply center A distribution point for expendable maintenance and administrative supplies (LOG)

SEMA Special electronics mission aircraft (AVN)

semiactive homing guidance Illumination of a target by a source separate from a missile that follows the beam's path (WPN)

semiautomatic Gun that automatically chambers the next round but requires a separate trigger pull to fire each round (WPN)

semifixed ammunition Ammunition in which the cartridge case and projectile are not permanently attached (ORD)

semimobile unit An element with insufficient transportation capability to move itself in one trip (LOG)

sensitive compartmented information (SCI) Classified intelligence product to which access must be specially restricted because of its critical nature (INTEL)

sentry Soldier detailed to maintain a watch against any threat to his or her unit (TAC)

separate Designation for a unit that is not part of a larger command to which it would normally be assigned, e.g., a company unassigned to a battalion (ADMIN)

separated ammunition Round that is divided into projectile and propelling charge components (ORD)

separation The release of a soldier from the Army for any reason, i.e., discharge, retirement, resignation, dropped from rolls, or death (ADMIN)

sequence circuit Mine detonating device calling for more than one influence in order for detonation (ORD)

SERE Survival, evasion, resistance, and escape (TNG)

sergeant An enlisted, noncommissioned officer; pay grade E-5 (PERS)

sergeant first class An enlisted, noncommissioned officer; pay grade E-7 (PERS)

sergeant major A senior enlisted, noncommissioned officer; pay grade E-9 (PERS)

serious criminal offense Any legal infraction for which the potential punishment includes the potential for a year or more of incarceration (LAW)

serviceability standards Categories for materiel issuability: Serviceable Group A, ready for issue; Serviceable Group B, ready for issue with minor repair; Unserviceable Group C, economically repairable; and Unserviceable Group D, uneconomically repairable (LOG)

serviceable Usable for the purpose intended (LOG)

service ammunition Ammunition intended for combat use rather than training (TAC)

service cap A round dress cap with a visor (PERS)

service club Support facility that provides a social and recreational program for off-duty enlisted personnel (LOG)

service message Traffic regarding handling, facilities, or circuit conditions (COMMS)

service ribbon Small, multicolored bars worn in lieu of full-sized decorations and service medals on dress uniforms (PERS)

service star Device worn point up on service ribbons to denote additional awards; one silver star equals five bronze stars; silver stars are worn to the wearer's right of bronze stars and to the left of arrowheads (PERS)

service stripe Cloth strip representing three years of enlisted service; worn on the left sleeve of dress uniforms (PERS)

service uniform A uniform for routine duty; a work uniform as opposed to a dress uniform (PERS)

setback The relative rearward motion of free-moving components of a projectile (ORD)

settling rounds Rounds fired to firmly set a large gun's base plate in the ground (ARTY)

severe damage A degree of destruction that permanently prevents the use of equipment or facilities (TAC)

SF **1.** Special forces (ADMIN); **2.** Standard form (ADMIN)

SFC Sergeant first class (PERS)

SFF Self-forging fragmentation (ORD)

SFG Special forces group (ADMIN)

SFOB Special Forces Operational Base (ADMIN)

SFOD Special Forces Operational Detachment (ADMIN)

SGLI Servicemen's Group Life Insurance (ADMIN)

SGM Sergeant major (PERS)

SGT Sergeant (PERS)

shadow factor A formula used in image interpretation to determine the height of objects based on the length of the shadow they cast compared to latitude and time of day (INTEL)

shallow fording The crossing of a water obstacle by a vehicle with its wheels in contact with the ground and without the need for a special water-proofing kit (TAC)

shaped charge A munition designed to project its explosive power in a controlled direction (ORD)

sharpshooter Mid-level qualification for accurate delivery of fire with an aimed weapon; above marksman and below expert (PERS)

sheaf Fire from two or more guns to produce a desired pattern of bursts (ARTY)

shell A hollow projectile to be filled with explosive or other material (ORD)

shelter half Man-portable half of a two-soldier tent (INF)

Sheridan The M551, a lightly armored, air-droppable tracked vehicle that functions as the principal assault weapon of airborne units (ABN)

shifting fire Fire delivered at constant range but varying deflections (ARTY)

shillealagh A tank-mounted, antiarmor missile system (WPN)

shock Life-threatening condition that can result from any major injury to the body; symptoms include confusion, blotchy or clammy skin, nervousness, thirst, vomiting, and increased breathing and pulse rate; treatment—make comfortable, loosen clothing, lie casualty down and elevate his or her feet, prevent chilling or overheating, reassure casualty, seek medical aid, do not provide food or drink (MED)

SHORAD Short-range air defense (AD)

short **1.** Having little time left on an enlistment or assignment (COLL); **2.** Fire of insufficient range to impact the intended target (ARTY)

short-range ballistic missile A missile with an effective range of up to 600 miles (WPN)

short-range radar Target-detection equipment capable of resolving a one-square-meter target at a range of 50 to 150 miles (INTEL)

short-tour area A locale overseas where duty assignments are less than 36 months with dependents or 24 months without dependents (ADMIN)

shot group Pattern formed on a target by impacting rounds from a weapon using one sight adjustment (TNG)

shoulder arms A two-part command preceded by "right" or "left" directing soldiers to place their weapons on the indicated shoulder (DC)

shoulder cord Loop of blue, interlocking square knots worn around the right shoulder by combat or expert infantrymen (INF)

showdown inspection Direct physical examination and inventory of clothing and equipment in the hands of soldiers to ensure their preparedness (ADMIN)

shrapnel The destructive fragments resulting from the explosion of a bomb or projectile (ORD)

shuttle marching Movement of troops via a combination of riding in available vehicles and walking alongside them (LOG)

SI Special intelligence (INTEL)

sick call The regular assembly of ill and/or injured soldiers for diagnosis and treatment (ADMIN)

side step Twelve-inch lateral pace without turning by an individual in formation (DC)

SIDPERS Standard Installation/Division Personnel System (ADMIN)

siege The encirclement, blockade, and bombardment of a fortified place (TAC)

SIGCEN Signal center (COMMS)

SIGINT Signals intelligence (INTEL)

signal A transmitted electronic impulse, or any means of conveying information between widely separated points, e.g., visual or acoustical (COMMS)

signal center A facility housing combinations of tactical and nontactical terminal and switching systems (COMMS)

signal communications Any means of message transmission other than direct conversation or courier (COMMS)

Signal Corps The combat support branch and personnel responsible for establishing and maintaining communications networks (ADMIN)

signals intelligence Intelligence derived from all forms of communicated information, including communications intelligence, electronics intelligence, and telemetry intelligence (EW)

signature Any unique indicator of the presence of certain materiel or troops (INTEL)

signature block The name, rank, and position of the sender of a communication (ADMIN)

signature equipment Weapons or materiel that indicate the presence of a unit known to possess such equipment (INTEL)

sign off To terminate a transmission (COMMS)

SIGSEC Signal security (COMMS)

silhouetting Targetting, especially at night, by attempting to see the target against the sky rather than the ground (TAC)

Silver Star Decoration for heroism involving risk of life in action against an enemy of the United States (PERS)

simplex A circuit that permits communication in only one direction at a time; half duplex (COMMS)

SIMU Suspended from issue, movement, and use (LOG)

simulative electronic deception (SED) The generation and broadcast of electromagnetic emissions meant to resemble those normally emitted by friendly units for the purpose of misleading the enemy (TAC)

SINCGARS Single-channel ground and airborne radio subsystem (COMMS)

single lane A cleared path through a minefield wide enough (eight meters) to accommodate one-way vehicular traffic (TAC)

single member sponsor An unmarried or legally separated soldier who has custody of or responsibility for dependent family members (ADMIN)

single section charge A propelling charge that cannot be broken down in attempting to adjust for range (ARTY)

SIOP Single Integrated Operational Plan (STRAT)

SIR Special information requirement (INTEL)

site The sum of the angle of site and the complementary angle of site that compensates for a target not at the same altitude as the battery (ARTY)

SITREP Situation report (INTEL)

SJA Staff judge advocate (ADMIN)

skill level A degree of expertise within an occupational specialty (TNG)

skip zone An area between the end of a transmitter's ground wave and the reflected sky wave where no signal is received (COMMS)

skirmish line A line of dismounted soldiers at staggered, extended intervals (INF)

skirt Light, protective armoring that hangs partially over a tank's track/suspension system (ARM)

SKO Sets, kits, and outfits (LOG)

sky wave The component of a radiated electromagnetic signal that is refracted off the ionosphere and returns to earth (COMMS)

SL safety level (LOG)

slant range Straight-line distance between two points, including difference in elevation (WPN)

SLAR Side-looking airborne radar (INTEL)

sleigh The part of a gun carriage that supports the barrel and the recoil mechanism (ARTY)

slew To traverse a gun (ARTY)

slice A portion of a unit's assets available for loan to other commands (ADMIN)

slick An aircraft without armaments or other optional equipment—one that is stripped for transport use (AVN)

sling 1. Light cargo carrier attached to the bottom of a helicopter (AVN); 2. A carrying strap (INF)

sling arms A two-part command directing troops in formation to place their weapons on their shoulders by using the carrying strap (DC)

slip Maneuver of a parachute canopy to adjust lateral drift or rate of descent (ABN)

slope The angle formed between the ground's surface and horizontal (TAC)

SLUFAE Surface-launched unit, fuel-air explosive (ORD)

slurry A mixture of water and dry decontaminant (NBC)

small arms Guns of a caliber up to and including 20 millimeters (WPN)

small-scale map A map at a scale of 1:600,000 or smaller (INTEL)

smart weapon Missile or other munition with the ability to identify a target and seek it out (WPN)

SMCT Soldiers' Manual of Common Tasks (TNG)

SMOS Secondary military occupational specialty (ADMIN)

SNAFU Situation normal—all fouled up (COLL)

snap link Steel connecting ring used primarily to slick along ropes while rappelling (INF)

sniper A rifleman who fires from a concealed position (TAC)

sniperscope A night observation device combined with a rifle (WPN)

snowbird To report early for a long-term duty assignment (COLL)

SO Stockage objectives (LOG)

SOCOM Special operations command (ADMIN)

SOF Special operations forces (ADMIN)

SOFA Status of Forces Agreement (STRAT)

SOL Out of luck (COLL)

Soldiers' and Sailors' Civil Relief Act Statute that suspends enforcement of certain civil liabilities against armed forces members (LAW)

Soldier's Medal Decoration made in recognition of heroism not involving actual combat (PERS)

soman (GD) A highly toxic nerve agent; decontaminate with STB slurry, bleach, DS 2, hot soapy water, or individual decontamination kit (NBC)

SOP Standard operating procedure (ADMIN)

SOR States of readiness (ADMIN)

sortie An air mission (AVN)

soundex code An encipherment of a soldier's name for the purpose of filing his or her personal financial information (ADMIN)

sound ranging The adjustment of gun or battery fire based on the time difference in the observation of impact and the sound of impact (ARTY)

source Any thing or person from which intelligence information is obtained (INTEL)

SP Self-propelled (WPN)

space-available Priority for transportation of personnel and materiel; will be transported only if excess capacity exists; also known as space-A (ADMIN)

space-imbalanced MOS Overseas requirements for a given military occupational specialty approaching or exceeding 55 percent of the Army-wide total (ADMIN)

spanning tray Device used to insert elements of separate loading ammunition into a cannon breech (ARTY)

spar bridge Expedient bridge built of lashed-together timbers (TAC)

spare A repair part (LOG)

sparrow team Assassination team consisting of civilian members such as adolescents or women who are able to approach their target because of their innocent appearance (OPFOR)

special **1.** Nuclear, biological, or chemical (WPN); **2.** Classified (INTEL)

special ammunition Ordnance requiring intensive handling, security, or management, such as nuclear rounds, missiles, and chemical rounds (WPN)

special branches Combat service support-type specialties that are managed separately from CSS branches—the Army Medical Department, chaplains, and the Judge Advocate General (ADMIN)

special category messages Communications related to projects or operations requiring narrower limitation of access than would be provided by normal security classifications (INTEL)

special court-martial A court of at least three members authorized to try any offense subject to the Uniform Code of Military Justice but generally limited to imposing punishment not in excess of six months of confinement and forfeiture of two thirds' pay for the six months (LAW)

special disbursing agent Any person authorized by the Secretary of the Army to disburse and maintain accountability of funds (ADMIN)

special forces The combat branch and units responsible for special reconnaissance, direct action, foreign internal defense, unconventional warfare, and counterterrorism missions (ADMIN)

specialist The rank of an enlisted soldier whose technical proficiency defines his or her position and relative status (PERS)

special operations Clandestine or covert missions against political, economic, psychological, or military targets in enemy territory (TAC)

special qualification digit A suffix to a military occupational specialty code indicating an additional degree of expertise required for a position (ADMIN)

special sheaf Any pattern of bursts other than parallel, converged, or open (ARTY)

special smoke A planned mission to deliberately build a screen of obscuring smoke (ARTY)

special staff Those officers not part of the general or personal staffs who assist and advise commanders in technical specialties, e.g., the provost marshal, transportation officer, and adjutant general (ADMIN)

special weapons Nonconventional weapons, nuclear chemical or biological munitions (ORD)

specified command A unit, normally composed of personnel a single service, that has a broad and continuing mission and is established by the authority of the President (ADMIN)

spectrum of war The range in intensity of conflict between nations, from cold war (characterized by espionage and terrorism) to limited war (proxy nations, guerrilla warfare) to general war (all-out direct combat) (MS)

speed-ring sight An aiming device with built-in lead computation for moving targets (WPN)

spetznaz Soviet special operations troops (OPFOR)

splinterproof shelter Defensive position that protects against small arms fire, grenades, and shrapnel from high-explosive shells up to three inches in size (TAC)

splinter village Garrison area of wooden, World War II era barracks (COLL)

SPM Security program manager (ADMIN)

spoiling attack An offensive maneuver designed to seriously impair or disrupt an enemy in the process of forming their own offensive (TAC)

sponson Equipment storage compartment on the hull or turret of a tank (ARM)

sponsor A soldier whose service membership entitles his or her dependent(s) to certain benefits (ADMIN)

spoon A cook; any mess hall personnel (COLL)

spot jamming Transmission of electromagnetic energy meant to disrupt a specific channel or frequency (TAC)

spotting The process of observing and adjusting artillery fire (ARTY)

spotting line Any reference line (e.g., gun-target, observer-target) used by an observer to adjust or correct fire (ARTY)

SQT Skill qualifications test (TNG)

squad An administrative subdivision of a platoon; the Army's smallest tactical unit (ADMIN)

squadron The basic administrative aviation unit (AVN)

square division An infantry division consisting of two brigades, each of which contains two regiments (MS)

squawk To broadcast a code that identifies an aircraft as friend or foe (AD)

squelch circuit A radio receiver control that reduces background noise in the absence of a signal (COMMS)

squib A pyrotechnic device used to set off a larger fuse or igniter (ORD)

squint An image interpreter (COLL)

SRB Selective reenlistment bonus (ADMIN)

SRO Senior ranking officer (ADMIN)

SSA Supply support activity (LOG)

SSAN Social Security account number (ADMIN)

SSG Staff sergeant (PERS)

SSN Social Security number (ADMIN)

SSO Special security officer (INTEL)

SSSC Self-service supply centers (LOG)

stabilized tour A CONUS duty assignment of a mandatory time length (ADMIN)

stack arms **1.** Temporary field or ceremonial storage of rifles (DC); **2.** To not work or not do one's duty (COLL)

staff A commander's group of specialized advisors (ADMIN)

staffing guide A reference manual that specifies the number and type of personnel needed to administer table of distribution units (ADMIN)

staff ride Visit to a field site to observe terrain and/or battlefield conditions; a field trip (TNG)

staff sergeant An enlisted noncommissioned officer in the pay grade E-6 (PERS)

stage To prepare troops and materiel for deployment (LOG)

STANAG Standardization agreement (ADMIN)

standard B Materiel that is being phased out but is still suitable for its intended use (LOG)

standard B ration Nonperishable meals suitable for use during field operations where refrigeration is not available (LOG)

standardization agreement (STANAG) The adoption by allied countries (i.e., NATO) of interchangeable or compatible materiel and supplies (LOG)

standard nomenclature Systematic method of naming materiel—the noun name of the item followed by modifiers in reverse of their conversational order; e.g., meal, ready-to-eat (LOG)

standard operating procedure (SOP) Sets of instructions to be followed under routine conditions in the absence of necessary exceptions in unusual circumstances (ADMIN)

standard requirements code A code that permits the adaptation to varying situations of the personnel and equipment authorized in tables of organization and equipment (ADMIN)

standard trajectory The path of a projectile under given conditions of weather, position, fuse, and charge (ARTY)

stand at ease Two-part command directing troops in formation to assume the position of parade rest but to continue looking at the person in charge of the formation (DC)

standby **1.** Command that soldiers or materiel be prepared for acton without further notice (ADMIN); **2.** Advice to a forward observer that bursts should occur in five seconds (ARTY)

standby reserve Reservists (active or inactive) who are retained voluntarily or involuntarily for possible active duty (ADMIN)

standby status Status of a unit with such severe personnel shortages that its equipment is placed in storage and the remaining personnel secure the area (ADMIN)

standby storage Holding space for materiel expected to be withdrawn within 90 days to three years (LOG)

stand down To relax an alert status (ADMIN)

standing order An order that remains in force until changed or cancelled (ADMIN)

stand off **1.** The net difference between the maximum effective range of weapons of opposing forces (WPN); **2.** In a shaped-charge round, the distance between the base of the liner and the point of impact (ORD)

STANFINS Standard Finance System (ADMIN)

STANO Surveillance target acquisition and night observation (WPN)

star A pyrotechnic round that burns as a single light (ORD)

star clusters Signaling and illumination rounds of various colors (ORD)

star gauge A device used to measure the bore of a gun (ARTY)

Starlifter The C-141, a fixed-wing cargo aircraft capable of airdrops (LOG)

Starlight scope The AN/PVS-2 electrically powered device for passive night vision (WPN)

STARPUBS Standard Army Publications System (ADMIN)

star shell Projectile used to illuminate a battlefield (ORD)

start point A well-defined position along a line of communications at which moving elements join under control of a movement commander (TAC)

statement of charges A bill payable by a soldier to the government for property lost, damaged, or destroyed while in his or her care (LOG)

state of alert Time until ready to fire at least one missile; standards are as follows: battle stations (30 seconds), 5 minutes, 15 minutes, 30 minutes, 1 hour, and 3 hours (STRAT)

static display A nonoperating exhibit (TNG)

static line Durable nylon strap connected between an aircraft and a parachute meant to pull the parachute from its pack to inflate it (ABN)

station time Time at which aircrew and jumpers are to be aboard an aircraft prepared for departure (ABN)

STAVKA Headquarters of the Soviet Supreme High Command's top wartime decision-making body (OPFOR)

status of forces agreement (SOFA) An agreement between allied nations governing the administration of foreign troops within a sovereign nation (ADMIN)

stay behind An agent who remains in an area under enemy control (INTEL)

STB Supertropical bleach (NBC)

steady on Continue traversing; a tank command (ARM)

step A full pace of 30 inches (DC)

sterile Untraceable; materiel (i.e., uniforms, equipment) that has been stripped of all indications of its identity and origin (UW)

stick 1. A squad of paratroopers who exit an aircraft on the same pass (ABN); 2. A pencil (COMMS)

stinger The FIM92A, a man-portable, shoulder-fixed missile for low-altitude air defense of forward area troops (AD)

stockade A military jail (COLL)

stockage objective Inventory level maintained to sustain current operations; the sum of operating and safety levels (LOG)

stock record account Management record used to document and control an activity's materiel inventory (LOG)

stop fire An emergency fire control order directing a battery to temporarily cease operations because of an unsafe condition within the unit (AD)

storm flag A nylon national flag measuring five feet by 9.5 feet that is flown during inclement weather (DC)

STP **1.** Soldier training publication (TNG); **2.** Separation transfer point (LOG)

strac A soldier possessing exemplary military bearing and appearance (COLL)

straddle trench Field-expedient latrine (TAC)

STRAF Strategic Army forces (STRAT)

strafing The delivery of automatic fire by aircraft against ground targets (TAC)

straggler A soldier who becomes unintentionally separated from his or her unit (TAC)

strategic reserve Uncommitted forces or materiel that are free to be deployed wherever needed (STRAT)

strategic warning Notification that enemy attack may be imminent (STRAT)

strategy The overall plan for utilization of political, economic, psychological, and military assets in support of national conflict (MS)

streamer A parachute canopy that deploys but does not inflate (ABN)

strike **1.** An offensive action (TAC); **2.** To take down or prepare for movement, especially a tent or flag (COLL)

striking force area That portion of a mobile defensive sector directly behind the forward defense area (TAC)

STRIKWARN Strike warning message (NBC)

strip alert Positioning of aircraft in a ready-to-deploy mode (AVN)

structure strength The level of a unit's personnel and equipment fill required to perform its sustained wartime mission (ADMIN)

STX Situational training exercise (TNG)

subcaliber ammunition Smaller than standard ammunition used to train on certain weapons without expending the full-size rounds they normally use (TNG)

subclasses of supply Categorizations of supply designed to enhance their manageability based on the following special factors: (a) transportation requirements, (b) packaging requirements, (c) storage and handling characteristics, (d) commodity characteristics, (e) intended usage, and (f) chemical/radiological sensitivity (LOG)

subcourse A block of specialized nonresident instruction (TNG)

subdued insignia Rank identification that is visible only at a limited range, to minimize the enemy's opportunity for observation (PERS)

submunitions Any device or bomb dispensed from a larger projectile or rocket (ORD)

subsistence Food (LOG)

substitution system An encryption system in which the plain text characters change but retain their original sequence or position (COMMS)

subversion Action that undermines the stability of a government (TAC)

summary court-martial A court of one officer limited to maximum punishment of one month of confinement, up to 45 days of hard labor, up to two months of restriction, and forfeiture of no more than two-thirds of a month's pay (LAW)

summit The high point of a round's trajectory (ARTY)

sundry funds Nonappropriated fund activities such as open messes that, although necessary, are required to be self-sustaining (ADMIN)

superelevation Elevation added to the aimed trajectory of a projectile to compensate for the effect of gravity on the projectile (ARTY)

superencipherment Compound encipherment; the encipherment of previously enciphered message text (COMMS)

superencryption Compound encryption; the encryption of previously encrypted message text (COMMS)

supernumerary An extra member assigned to a guard detail in case of the incapacitation of other guard personnel (ADMIN)

supertropical bleach (STB) A decontaminant that neutralizes chemical agents (NBC)

supplementary target A secondary target, on which fire is delivered after primary targets are destroyed or not subject to effective fire (ARTY)

supply credit The allocation of a quantity of materiel to a unit, to be drawn at the commander's discretion (LOG)

supply economy The conservation of and care for all Army materiel in support of operations (LOG)

supply support activity (SSA) A unit whose mission includes materiel or maintenance assistance to other units and that is assigned a Department of Defense activity address code (LOG)

support To help or sustain another force (MS)

supporting attack An offensive operation that aids a main attack by attempting to deceive, destroy, or interdict the enemy, control terrain key to the main attack, or force early commitment of the enemy reserve (TAC)

supporting distance Maximum separation up to which tactically deployed elements can aid each other through fire and maneuver (TAC)

supporting range The distance through which effective fire can be delivered with available weapons (TAC)

support unit A separate command with the specific mission of assisting other commands (ADMIN)

suppressive fire Fire designed to degrade the performance of an enemy weapons system (TAC)

surface lines Telephone or telegraph wire laid on the ground in a hasty operation until it can be permanently installed (COMMS)

surface mission The employment of air defense weapons against ground or naval targets (AD)

survey control station A position of known location and elevation from which other positions can be precisely determined (ADMIN)

surveying officer The officer assigned to investigate the loss, damage, or destruction of government property and determine responsibility and liability (LOG)

suspect battery Enemy unit whose existence is confirmed but whose exact position is not known (INTEL)

suspense A deadline (ADMIN)

suspension of arms Brief truce arranged for purposes such as attending to casualties or the dead (LAW)

sustainability The ability to maintain and support combat operations in support of national objectives (MS)

sustained rate of fire The rate at which a weapon can operate without overheating (WPN)

swallow An espionage agent who utilizes sexual entrapment (OPFOR)

sweep **1.** To rapidly scan a range of electromagnetic frequencies (EW); **2.** To cover an area with fire by rapid changes in a weapon's deflection (TAC)

sweep jamming Transmission of electromagnetic energy meant to disrupt enemy communications by cycling through the entire spectrum and locking on targeted signals (EW)

swimming capability The ability of a vehicle to cross a water obstacle without being in contact with the bottom (TAC)

swimming device Equipment attached to a vehicle for the purpose of floating it across water obstacles (TAC)

swinging traverse fire A pattern of impacting rounds of constant range but direction that varies from point to point (TAC)

swiss seat A personal rope used in rappelling and air assault deployment (INF)

switch position A defensive position connecting successive main lines of defense (TAC)

SWO Staff weather officer (PERS)

sympathetic detonation The explosion of a charge due to the nearby explosion of another charge (ORD)

synchronization The coordinated violent application of maximum combat power at a critical time and place (MS)

synthesis The examination and combination of information and intelligence for final interpretation (INTEL)

system indicator Symbols or characters that designate a specific cryptosystem (COMMS)

T

TA Target acquisition (ARTY)

TAACOM Theater Army area command (ADMIN)

TAADS The Army Authorization Documents System (LOG)

TAB Target acquisition battalion/battery (ARTY)

Tabard Banner attached to a bugle (DC)

table of distribution and allowance A document that authorizes materiel for unit or mission requirements for which there is no appropriate table of organization and equipment (LOG)

table of organization and equipment A document that authorizes a unit's formation, personnel, and equipment and prescribes its mission (ADMIN)

taboo frequency A portion of the radio communications spectrum that is not to be jammed because it is being used for friendly communications (EW)

tabun (GA) Highly toxic nerve agent; decontaminate with STB slurry, bleach, DS 2, hot soapy water, or individual decontamination kit (NBC)

tac Tactical (COLL)

TACFIRE Tactical fire direction system (ARTY)

tactical Pertaining to combat (MS)

tactical air coordinator An officer who directs, from an air platform, the employment of close air support (AVN)

tactical air support Direct combat assistance to ground forces by aircraft (TAC)

tactical air transport operations The movement of personnel and cargo within an active theater of military operations (AVN)

tactical column Movement organized for speed, but in a configuration that lends itself to rapid assumption of a defensive formation in the unlikely event of enemy contact (TAC)

tactical minefield Large, relatively permanent emplacements of mines designed to influence the course of battle by restricting the enemy's ability to maneuver (TAC)

tactical operations center (TOC) The headquarters within which general and special staffs plan the conduct and support of current combat maneuvers (ADMIN)

tactical plan A combat operation plan omitting provision for support functions (ADMIN)

tactical reserve Forces or materiel held out of its battle, under the control of or dedicated to a specific local command (STRAT)

tactical standard mobility The capacity for occasional cross-country movement (LOG)

tactical vehicle Any noncommercial truck or transporter ruggedized for combat use (LOG)

tactical warning Notification of the initiation of hostilities or occurrence of a specific hostile action (INTEL)

tactics The translation, at levels corps and below, of potential combat power into successful battles or engagements (MS)

TAFFS The Army Functional Files System (ADMIN)

TAG The Adjutant General (PERS)

TAMMS The Army maintenance management system (LOG)

tank, main battle A tracked, armored, highly mobile vehicle designed to provide fire support to offensive operations (ARM)

tank tables Series of gunnery and tactical tasks used in the training and evaluation of armor crews (TNG)

taps Bugle call played at the end of the duty day and at military funerals (DC)

target acquisition The detection, identification, and location of enemy assets in enough detail to permit effective use of weapons against them (TAC)

target analysis The examination of potential enemy assets to determine their significance and consequent priority, and the determination of which weapon will produce the desired level of damage (TAC)

target discrimination The ability of a system to identify a single target when multiple targets are present (WPN)

target folder An assembly of all information required for execution of action against a particular enemy position (INTEL)

target grid Device for equating observers' target corrections to corrections referenced to the gun target line (ARTY)

targeting The process of identifying and locating enemy assets and matching the appropriate friendly assets against them in support of operational requirements (TAC)

target of opportunity A target that appears during an operation and for which fire has not been previously planned (TAC)

task A clearly defined, measurable activity that contributes to accomplishment of a mission by either an individual or a unit (TNG)

task force The temporary combination of assets under a single commander in order to carry out a specific mission (ADMIN)

tasking The translation of military requirements into orders and their allocation and communication among the appropriate units (ADMIN)

task organization The tailoring of a force by nonstandard combination of units for efficient execution of a specific mission (ADMIN)

TBD To be determined (COLL)

TBM Tactical ballistic missile (AD)

TBP To be published (COLL)

TC **1.** Training circular (TNG); **2.** Transportation Corps (ADMIN)

TCAE Technical control and analysis element (INTEL)

TCN Transportation control number (LOG)

TD Tank division (ADMIN)

TDA Tables of distribution and allowance (ADMIN)

TDY Temporary duty (ADMIN)

Teampack The AN/MSQ0103A, a fully tracked vehicle capable of detection, interception, and direction-finding of noncommunications emitters (EW)

TEC Training extension course (TNG)

technical bulletin An update on equipment maintenance (MAINT)

technical manual A publication that provides detailed explanation of doctrine or equipment beyond that found in field manuals (ADMIN)

technical means A way of gathering of intelligence by other than human agents (INTEL)

technician A full-time civilian employee assigned in support of, and who is normally a member of, a National Guard or Reserve unit (PERS)

TELAR Transporter, erector, launcher, and radar (AD)

telecommunication Any long-distance transmission of information (COMMS)

telegraphy Communication in the form of on/off, long/short binary signals, as with Morse code (COMM)

telemetry Remote measuring; the sensing of any physical characteristic and the automatic transmission of such measurements elsewhere for interpretation (INTEL)

telephony High quality voice communication via land-line circuitry (COMMS)

teleprocessing The combination of data processing and telecommunications (COMMS)

tempest Terminal electromagnetic pulse escape safeguard technique—a program of controlling compromising emissions by data processing equipment (INTEL)

template A graphic representation of the doctrinal deployment of forces (INTEL)

temporary disability retired list A list of soldiers who are physically unable to continue on active duty and must undergo evaluation of their condition periodically (ADMIN)

terminal phase The final portion of a missile's flight path, from atmosphere reentry through impact (AD)

terminal service company A Transportation Corps unit capable of loading and unloading passengers and cargo from ships, trains, trucks, or aircraft (LOG)

terrain The land and everything—natural or man-made—affixed to it (TAC)

terrain analysis The visualization of an area of ground based on its depiction on a map (INTEL)

terrain evaluation An examination of the limiting factors of terrain in an area of potential military operations, on both friendly and enemy courses of action (INTEL)

terrain flights Low-altitude aviation operations that follow the earth's contours to minimize the aircraft's vulnerability to ground observation and fire (AVN)

terrain masking The reduction of visibility by certain terrain features (TAC)

terrain return Undesired reflection of radar signals by the ground or insignificant terrain features (INTEL)

terrorism Isolated acts of violence or destruction directed at an entire society for the purpose of undermining confidence in the government's ability to maintain order (UW)

test piece Any gun in a battery other than the reference piece (ARTY)

TEWT Tactical exercises without troops (TNG)

TFT Tabular firing table (ARTY)

theater A large geographical area outside the continental United States for which military responsibility has been assigned as a unified or specified command (ADMIN)

thermal effect The heat energy resulting from a nuclear detonation (NBC)

thermal imagery Visual representation of temperature differences; a system enabling "night vision" (INTEL)

thermate A filling for incendiary munitions that burns at approximately 4,300 degrees Fahrenheit (ORD)

thickened fuel Gelatinized gasoline used in bombs and flamethrowers (WPN)

third world nation A nonindustrial, underdeveloped, and/or impoverished nation (STRAT)

Threatcon Red Highest level of potential terrorist activity, reflecting an imminent, specific threat against personnel or facilities (UW)

Threatcon White Lowest level of potential terrorist activity, representing a nonspecific threat within a general geographic area (UW)

Threatcon Yellow An intermediate level of potential terrorist activity, reflecting a specific threat within a geographic area (UW)

throughout distribution Delivery of materiel directly to using units, bypassing intermediate supply echelons (LOG)

TIARA Tactical intelligence and related activities (INTEL)

time of attack The time at which a leading element is to cross the line of departure in an offensive operation (TAC)

time of recognition The time at which an existing signal is identified as hostile or significant (EW)

time on target Planned time of arrival at an objective (TAC)

TISA Troop issue subsistence activity (LOG)

TM Technical manual (LOG)

TMDE Test, measurement, and diagnostic equipment (LOG)

TMP Transportation motor pool (LOG)

TO Theater of operations (ADMIN)

TOC Tactical operations center (ADMIN)

TOE Table of organization and equipment (ADMIN)

toe popper A nonmetallic antipersonnel mine designed to incapacitate rather than to kill (ORD)

tolerance dose Radioactive dose to which an individual can be exposed with negligible results (NBC)

Tomahawk An air-, land-, or sea-launched cruise missile (WPN)

Top First sergeant (COLL)

topographic map A map that depicts the elevation as well as the normal depiction of horizontal distribution (NAV)

top secret A classification for national security material requiring the highest degree of protection, the unauthorized disclosure of which could cause exceptionally grave damage to national security (INTEL)

TOT Time on target (ADMIN)

Total Army All Army personnel assets, including all Reserve components, retired soldiers, and DA civilian workers (ADMIN)

total dosage attack A nuclear or chemical attack of limited yield, designed to merely build up individuals' level of toxicity through a surprise burst or one that does not appear to warrant full defensive measures (TAC)

total materiel assets The inventory level of an item, including on-hand and funded procurement quantities (LOG)

total mobilization Activation of all Reserve units, available personnel, and generation of additional personnel to meet the national security requirements resulting from an external threat (STRAT)

tour A period of assignment to a duty station (ADMIN)

tourniquet Any device used to compress a blood vessel to control bleeding (MED)

TOW Tube-launched, optically tracked, wire-guided; heavy antitank missile (WPN)

TPFDD Time-phased force and deployment data (ADMIN)

TPFDL Time-phased force and deployment list (ADMIN)

TPU Troop program unit (ADMIN)

TR Transportation request (LOG)

tracer Bullet chemically treated to make its path visible in darkness (ORD)

track **1.** An endless belt that provides locomotion for an armored vehicle (ARM); **2.** The path over the earth's surface of an aircraft (AD); **3.** To display, record, or aim at successive positions of a moving object (TAC)

track telling The dissemination of air surveillance information between commands or facilities (AD)

TRADOC United States Army Training and Doctrine Command (ADMIN)

trafficability The ability of terrain to bear continuous vehicular traffic (TAC)

traffic flow security Constant transmission on a communications circuit to avoid tipping off the transmission of a valid message (COMMS)

traffic management The control of freight and passenger transportation services (LOG)

Trailblazer The AN/TSQ-114A, a fully tracked, tactical direction-finding and interception system (EW)

trains Series of logistics/materiel support elements (LOG)

transattack period The time from the beginning of a nuclear exchange to its termination (STRAT)

TRANSCOM Transportation command (ADMIN)

transfer hazard Liquid or solid contaminants (NBC)

transfer loader A wheeled vehicle with a movable platform used to move cargo among aircraft, ships, and vehicles (LOG)

transmission factor The numerical percentage of radiation that a given material permits to pass through it (NBC)

transponder A device on an aircraft that identifies it as friendly by receiving an electronic challenge and answering with a coded response (AVN)

Transportation Corps The combat service support branch and personnel who manage cargo and personnel-moving assets (ADMIN)

transshipment point A site prepared for the transfer of cargo between vehicles (LOG)

traveling Form of movement in which one element follows the other at a normal interval; used when speed is important and contact with the enemy unlikely (TAC)

traveling overwatch Form of movement in which the elements are spaced to prevent their simultaneous engagement by the enemy but close enough to enable them to support each other if necessary; used when speed is desirable but enemy contact possible (TAC)

traverse To rotate a gun horizontally in its mount (WPN)

traversing fire A pattern of impacting rounds that is of constant range but varying direction (WPN)

treason The betrayal of one's country by giving aid and comfort to its enemies (LAW)

TREE Transient radiation effect on electronics (COMMS)

triage The immediate sorting and evacuation of casualties, with the objective of allocating limited medical resources to provide optimum benefit to all (MED)

triangulation station A point for which the exact location has been determined and that therefore can be used as the basis for other calculations of position (NAV)

trick A work period; a shift (COLL)

triple canopy forest A densely wooded area with three distinct levels of tree tops, typically at 5 to 10, 20 to 25, and 35 to 45 meters (TAC)

trip ticket Form that authorizes possession and operation of a vehicle for a specific mission (LOG)

TRMF Theater readiness monitoring facility (ADMIN)

troop A cavalry unit at a command level below a squadron; equivalent to a company or battery (ADMIN)

troop issue subsistence activity (TISA) An installation's food warehouse (LOG)

troop left (or right) *See* battery left (or right) (ARTY)

troop program unit A selected Reserve unit that has been organized and authorized by a table of distribution and allowances or table of organization and equipment (ADMIN)

troop space cargo Personal equipment stowed so that it is accessible for use by owning soldiers in transit (LOG)

trophy of war Captured enemy materiel, the retention of which is not prohibited by law or regulation (LAW)

tropicalization Preparation of equipment for deployment to tropical climates by sealing it against humidity and the growth of fungus (LOG)

tropospheric scatter Irregular propagation of radio waves due to the effect of the lowest atmospheric level (COMMS)

true north The direction from any position toward the geographic North Pole (NAV)

TTY Teletypewriter (COMMS)

tube The barrel of a large-caliber gun (ARTY)

turning movement An offensive maneuver that bypasses an enemy's frontal defenses and targets objectives toward his or her rear, forcing the enemy to turn and face that threat (TAC)

turning point A bend of no more than 45 degrees in a minefield baseline (ORD)

turret An armored, usually rotating tower-like structure atop a tank (ARM)

TVA Target value analysis (ARTY)

TVD Soviet theater of military operations (OPFOR)

TWI Training with industry (TNG)

twilight Those periods of incomplete lightness or darkness when the sun is just below the horizon (ADMIN)

twin sideband Method of radio transmission that suppresses the carrier wave and passes the information via the upper and lower sidebands (COMMS)

TWOS Total warrant officer system (ADMIN)

TWX Teletypewriter message (COMMS)

type unit A command that has been assigned a five-digit, alphanumeric unit-type code (ADMIN)

U

U Utility (ADMIN)

UCMJ Uniform Code of Military Justice (LAW)

UH-1H The Iroquois, or Huey, helicopter; a light, rotary-wing, multipurpose aircraft (AVN)

UI **1.** Unidentified (COLL); **2.** Unit of issue (LOG)

UIC Unit identification code (ADMIN)

ullage Space left in a closed tank to allow for thermal expansion of fluid within it (LOG)

UMMIPS Uniform Materiel Movement and Issue Priority System (LOG)

UMR Unit manning report (ADMIN)

unaccompanied A tour of duty that does not permit a soldier to bring dependents (ADMIN)

unavoidable absence An unauthorized absence that occurs through no fault of the soldier or the government (ADMIN)

unconventional warfare Continuing special operations, guerrilla warfare, subversion, and/or sabotage in support of limited military, political, or economic objectives in denied areas (MS)

uncover **1.** Command of execution directing troops in formation to stagger their ranks (DC); **2.** To remove one's headgear (COLL)

UND Urgency of need designator (LOG)

underground A covert paramilitary network designed to operate in areas denied to guerrilla forces (UW)

undesirable discharge Characterization of a soldier's term of service when he or she is determined to be unsuitable for service because of misconduct, incapacity, or security reasons (ADMIN)

Unified Action Armed Forces A publication that presents the principles and doctrine governing the combined operation of two or more U.S. services (ADMIN)

unified command A unit with a broad continuing mission, composed of components of two or more services; established through the authority of the President (ADMIN)

Uniform Code of Military Justice (UCMJ) The legal system applicable to soldiers on active duty (LAW)

Uniform Materiel Movement and Issue Priority System (UMMIPS) A management system designed to ensure the adequate and appropriate distribution of supplies during war or peacetime (LOG)

unilateral frequency distribution A tabulation of how often individual cipher characters appear in an encrypted message (COMMS)

unit categories A citation of an entire unit's action; eligible to be worn by members who participated in the action (ADMIN)

United States Armed Forces Institute A DOD-sponsored agency that makes available professional and continuing education to off-duty soldiers (TNG)

United States Army The Regular Army, National Guard of the United States, and the Army Reserve (ADMIN)

United States Army Reserve A federal force of units and individuals that train part-time in anticipation of activation to support active duty units during a national emergency (ADMIN)

unit identification code (UIC) A unique, six-character alphanumeric designator that specifies each unit in the armed forces (ADMIN)

unit of issue (UI) The method of quantifying an item of materiel for management purposes (e.g., boots—pairs; trucks—each; wire—feet; sand—tons) (LOG)

unit train Combat service support functions under the direct control of the supported commander (LOG)

unit training Advanced, collective training that follows individual training; normally conducted under field conditions (TNG)

unit type code A five-character alphanumeric designator that identifies specific categories of units (ADMIN)

unity of command The synchronization of military effort under one responsible commander (MS)

universal mission load Those items of class II and IV materiel that are applied against a unit's combat mission, based on their table of organization and equipment, and subject to adjustment based on the specific mission (LOG)

universal time 0 The calculation of time based on the rotation of the earth, uncorrected for polar motion or seasonal variation (ADMIN)

universal time 1 Universal time 0, corrected for the earth's polar precession (ADMIN)

universal time 2 Universal time 1, corrected for variations in the earth's rotation (ADMIN)

unserviceable Materiel that is worn, damaged, or obsolete to the point where it cannot be used for its intended purpose (LOG)

unsupported meals Meals issued in excess of the authorized number (LOG)

unwarned exposed The vulnerability of friendly forces to the effects of a nuclear detonation when no warning is given; troops are assumed to be standing at the time of the detonation and prone by the time the blast wave hits (NBC)

urgency-of-need designator A letter code within the materiel issue priority system that helps determine a unit's immediacy of need for materiel, based on its deployment status (LOG)

urgent priority The second highest precedence of mission request; below emergency and above ordinary—used in cases such as enemy artillery falling on friendly forces or an imminent breakthrough (ADMIN)

usable rate of fire The practical speed at which a gun will operate, as opposed to the theoretical maximum rate of fire (WPN)

USADIP U.S. Army deserter information point (ADMIN)

USAREUR United States Army, Europe (ADMIN)

USG United States government (ADMIN)

USMA United States Military Academy (TNG)

USMTF United States message text format (COMMS)

USR Unit status report (ADMIN)

USSID United States signal intelligence directive (EW)

USSS United States SIGINT System (INTEL)

UTA Unit Training Assembly (TNG)

UTES Unit training equipment sites (LOG)

UTM Universal transverse mercator (NAV)

UW Unconventional warfare (MS)

UWOA Unconventional Warfare Operations Area (UW)

VA Department of Veterans Affairs (ADMIN)

validated positions Officer assignments requiring graduate-level training (ADMIN)

variant One of two or more cipher characters with the same plain text equivalent (COMMS)

V device An appurtenance worn in conjunction with certain service ribbons to denote valor (PERS)

vector **1.** An aircraft directional heading, in degrees (AVN); **2.** A disease-bearing microorganism or insect (MED)

vertical envelopment An offensive maneuver of air-delivered troops against the enemy rear, to encircle them and impede their withdrawal (TAC)

very high nuclear yield Nuclear weapon burst that generates over 500 kilotons of blast energy (WPN)

very long range radar Target-detection equipment capable of resolving a one-square-meter target at a range in excess of 600 miles (INTEL)

very low nuclear yield Nuclear weapon burst that generates less than one kiloton of blast energy (WPN)

very-short-range radar Target-detection equipment capable of resolving a one-square-meter target at a range of up to 50 miles (INTEL)

vetting The investigation of a source or agent to determine his or her loyalty and reliability (INTEL)

VHA Variable housing allowance (ADMIN)

VHF Very high frequency (COMMS)

vic Vicinity (TAC)

VINSON A family of secure communications devices (COMMS)

VOCO Verbal order of commanding officer (ADMIN)

vocoder A device that digitizes speech, enabling it to be broadcast at a reduced band width (COMMS)

volcano A multiple delivery mine system; a means for emplacing fields of various types of mines (ORD)

volley A burst of a specified number of rounds from the guns in a battery without an attempt to coordinate their timing (ARTY)

voluntary training Reservist participation in a nonpay status, for retirement points only (TNG)

voluntary training unit An entire reserve unit in which members drill for retirement points only (ADMIN)

VRB Variable reenlistment bonus (ADMIN)

V series A group of odorless, colorless toxic nerve agents (NBC)

Vulcan The M163/M167, a light, self-propelled, tracked antiaircraft gun system (AD)

vulnerability The degree to which an asset can survive on the battlefield (AD)

VX A highly toxic chemical nerve agent; decontaminate with hot soapy water, STB slurry, DS 2, or individual decontamination kit (NBC)

WAD Weapons alert designator (AD)

wanigan Sled-mounted shelter, for use in arctic regions (TAC)

warehouse refusal Advice by a supply support activity that a requisition for a particular item cannot be fulfilled (LOG)

warhead The explosive package section of a missile or rocket (ORD)

WARM Wartime reserve mode (EW)

WARMAPS Wartime manpower planning system (ADMIN)

warned exposed The vulnerability of friendly forces to the effects of a nuclear detonation when a warning allows only enough time to get prone and attempt to cover exposed skin (NBC)

warned protected The vulnerability of friendly forces to the effects of a nuclear detonation when a warning allows time to find shelter in a vehicle or improved position (NBC)

warning net A circuit specifically dedicated to transmission of information on a threatening enemy maneuver (COMMS)

warning order Brief, unformatted message that alerts subordinate units of an impending mission (TAC)

warrant officer A technician of such highly specialized skill as to merit officer rank without the general command responsibilities normally assigned to commissioned officers (PERS)

war reserves Those stocks of materiel maintained to sustain the early stages of a conflict until resupply can be established (LOG)

WARS Worldwide Ammunition Reporting System (ORD)

Warthog The A-10, a close-air support, antiarmor, fixed-wing aircraft (AVN)

wartime reserve mode The intentional suppression of friendly electromagnetic emissions in an emergency or immediately prior to an operation, in order to limit the enemy's ability to derive intelligence from them (EW)

WASPM Wide area side penetrator mine (ORD)

water buffalo A towed tanker used to deliver water to troops in the field (LOG)

wavelength The distance between successive peaks of radiated electromagnetic energy (COMMS)

waybill A document accompanying a materiel shipment that lists the consignor, consignee, point of origin, route, and final destination (LOG)

WB White bag (propellants) (ARTY)

WCS Weapons control status (AD)

weapon selector A circular scale used to depict the radius of nuclear damage on a map (NBC)

weapons free Weapons control status meaning that targets can be engaged until positively identified as friendly (AD)

weapons hold Weapons control status meaning that targets can be engaged only in self-defense (AD)

weapons tight Weapons control status meaning that only targets positively identified as enemy can be engaged (AD)

wear tables The correlation of gun tube use and resulting decreased muzzle velocity (ARTY)

weather intelligence Meteorological data forecasting and its impact on future combat operations (INTEL)

weather minimum The worst conditions under which aviation operations can be conducted (AVN)

web gear Field suspenders and pistol belt; tactical, individual, carrying equipment (TAC)

WESTCOM U.S. Army Western Command (ADMIN)

wet stowage Storage of large caliber ammunition in racks surrounded by nonflammable liquid to reduce the fire hazard (ORD)

WEZ Weapon engagement zone (AD)

white forces NATO designation for forces opposed to enemy forces within OPFOR exercises (INTEL)

whiteout Disorientation due to excessive snow and consequent loss of the horizon (TAC)

white phosphorus (WP) Chemical used in projectiles to produce smoke and an intense incendiary effect (ORD)

white propaganda Propaganda that is attributed to and acknowledged by its originator (UW)

WIA Wounded in action (ADMIN)

WILCO Proword meaning will comply (COMMS)

Wild Weasel The RF4-G, an aircraft specifically modified to locate and suppress enemy air defense systems (AVN)

willie peter White phosphorus (ORD)

windage Deviation in the trajectory of a round due to the effect of wind (WPN)

wind corrector A device that calculates the artillery adjustment to compensate for the effect of wind (ARTY)

wind drift The apparent shift in position of an object being sound ranged due to the effect of wind on the sound waves (ARTY)

wind dummy Anything dropped from an aircraft to gauge wind direction and strength (ABN)

window Time period of opportunity (ADMIN)

wind shear Layered air currents in direct contact with each other but moving in different directions (AVN)

wing The basic operational unit of air forces (ADMIN)

wire guided A projectile that receives direction in flight through cables trailing behind it (WPN)

wire head A command's forward limit of hardwire communications (COMMS)

wire roll An obstacle to the movement of wheeled vehicles consisting of a continuous steel spiral (TAC)

withdrawal The planned disengagement from contact with the enemy (TAC)

WLR Weapons locating radar (ARTY)

WO **1.** Warrant officer (PERS); **2.** Warning order (TAC)

WP White phosphorus (ORD)

WW Flash override message precedence (COMMS)

WWMCCS Worldwide Military Command and Control Systems (ADMIN)

WX **1.** Weather (INTEL); **2.** Simplex working (COMMS)

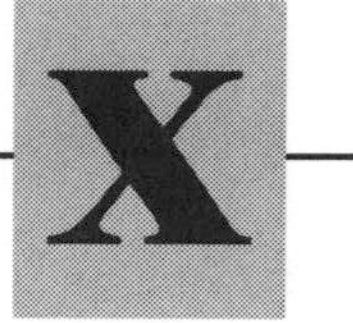

X Experimental (ADMIN)

XMITR Transmitter (COMMS)

XO Executive officer (ADMIN)

X-site An outdoor site for temporary storage of ammunition (LOG)

year group The annual period in which an officer was commissioned (PERS)

yellowleg A cavalry soldier (COLL)

Y-site An outdoor site, surrounded by earthen barricades, for temporary storage of ammunition (LOG)

Z

Z Zulu (COMMS)

Zen Operating signal for addressee to whom a message is sent by means other than electronically transmitted message (COMMS)

zero beat Signal used to tune the frequency calibration of radio sets (COMMS)

zero deflection Sight adjustment exactly parallel to a gun's bore axis (ARTY)

zero height of burst Fuse setting when impacting rounds result in an equal number of airs and grazes (ARTY)

zeroize To return machine cryptovariables to a basic, nonoperational setting (COMMS)

zero point The location of the center of the burst of a nuclear weapon (WPN)

ZI **1.** Flash interrupt message precedence (COMMS); **2.** Zone of the interior (STRAT)

zone I The roughly circular area surrounding ground zero of a nuclear burst requiring evacuation of personnel or maximum protective measures (NBC)

zone II The roughly circular area surrounding zone I of a nuclear burst requiring maximum protection for personnel (NBC)

zone III The roughly circular area surrounding zone II of a nuclear burst requiring minimum protection for personnel (NBC)

zone fire A series of rounds fired at a constant deflection but with varying elevation (ARTY)

zone of action A sector of responsibility assigned to a tactical unit, normally during an offensive operation (TAC)

zone of interior National territory outside the theater of operations (STRAT)

zone time system The division of the world into 24 segments 15 degrees wide for time-keeping purposes; beginning at the International Date Line and moving east, each zone is one hour later than the preceding one; *See* Table A-7 (ADMIN)

zulu time Greenwich mean time (ADMIN)

ZZ Flash message precedence (COMMS)

Reference Tables

TABLE A-1
Phonetic Alphabet

Oral communications are subject to misunderstanding. To minimize the possibility of misunderstanding spoken letters, prowords are used in place of letters that sound alike, according to the following table.

A	Alpha
B	Bravo
C	Charlie
D	Delta
E	Echo
F	Foxtrot
G	Golf
H	Hotel
I	India
J	Juliet
K	Kilo
L	Lima
M	Mike
N	November
O	Oscar
P	Papa
Q	Quebec
R	Romeo
S	Sierra
T	Tango
U	Uniform
V	Victor
W	Whiskey
X	X-ray
Y	Yankee
Z	Zulu

TABLE A-2
Morse Code

A	• —	N	— •
B	— • • •	O	— — —
C	— • — •	P	• — — •
D	— • •	Q	— — • —
E	•	R	• — •
F	• • — •	S	• • •
G	— — •	T	—
H	• • • •	U	• • —
I	• •	V	• • • —
J	• — — —	W	• — —
K	— • —	X	— • • —
L	• — • •	Y	— • — —
M	— —	Z	— — • •
1	• — — — —	6	— • • • •
2	• • — — —	7	— — • • •
3	• • • — —	8	— — — • •
4	• • • —	9	— — — — •
5	• • • • •	0	— — — — —

TABLE A-3
Q and Z Signals

Q and Z signals are brevity codes or prosigns that are widely understood by Morse communicators of all languages.

Q SIGNALS

Signal	Question	Answer, Advice, or Order
QAP	Shall I listen for you (or for ______) on ______ kHz (or MHz)?	Listen for me (or for ______) on ______ kHz (or MHz).
QCP	How is my tone?	Your tone is poor.
QRA	What is your call sign?	My call sign is ______.
ORG	What is my exact frequency (or that of ______)?	Your exact frequency (or that of ______) is ______ kHz (or MHz).
QRH	Does my frequency vary?	Your frequency varies.
QRK	What is my readability?	Your readability is ______. 1. Bad 2. Poor 3. Fair 4. Good 5. Excellent
QRL	Are you busy?	I am busy (or I am busy with ______). Please do not interfere.
QRM	Are you being interfered with?	I am being interfered with.
QRN	Are you troubled with static?	I am troubled with static.
QRO	Shall I increase power?	Increase power.
QRP	Shall I decrease power?	Decrease power.
QRQ	Shall I send faster?	Send faster (______ w.p.m.).
QRR	Are you ready for automatic operation?	I am ready for automatic operation. Send at ______ words per minute.
QRS	Shall I send more slowly?	Send more slowly (______ w.p.m.).
QRT	Shall I stop sending?	Stop sending.
QRU	Have you anything for me?	I have nothing for you.
QRV	Are you ready?	I am ready.
QRW	Shall I inform ______ that you are calling him on ______ kHz (or MHz)?	Please inform ______ that I am calling him on ______ kHz (or MHz).
QRX	When will you call again? Shall I close down until ______?	I will call you again at ______ hours.
QRZ	Who is calling me?	You are being called by ______.

TABLE A-3
(continued)

Q SIGNALS

Signal	Question	Answer, Advice, or Order
QSA	What is my signal strength?	Your signal strength is ______. 1. Scarcely perceptible 2. Weak 3. Fairly good 4. Good 5. Very good
QSB	Is my signal fading?	Your signal is fading.
QSL	Do you acknowledge receipt?	I acknowledge receipt.
QSO	Can you communicate with ______?	I can communicate with ______ directly (or by relay through ______).
QSU	Shall I send or reply on this frequency (or on ______)?	Send or reply on this frequency (or on ______).
QSV	Shall I send a series of V's for tuning?	Send a series of V's for tuning.
QSW	Will you send on this frequency (or on ______)?	I am going to send on this frequency (or on ______).
QSY	Shall I change to another frequency?	Change to another frequency (or to ______).
QTA	Shall I cancel message NR ______?	Cancel message NR ______.
QTC	How many messages do you have for me?	I have ______ messages for you.
QTH	What is your position?	My position is ______ latitude, ______ longitude.
QTR	What is the correct time?	The correct time is ______.

Z SIGNALS

Signal	Question	Answer, Advice, or Order
ZAL		Alter your wave length (frequency).
ZAN		We can receive absolutely nothing.
ZAP		Acknowledge please.
ZAR		Revert to automatic relay.
ZBD		Your signals are blurring badly.
ZBS		You have a blurring signal.
ZCK		Check keying.

TABLE A-3
(continued)

Z SIGNALS

Signal	Question	Answer, Advice, or Order
ZCO		Local receiving conditions poor; increase your power to the maximum.
ZCS		Cease sending.
ZCT		Send code twice.
ZCW	Are you in direct communications with ______?	
ZFB		Your signals are fading badly.
ZFS		Your signals are fading slightly.
ZHA	How are your conditions for auto reception?	
ZHC	How are your receiving conditions?	
ZMO		Stand by a moment.
ZNB		We do not get your breaks, so we send twice.
ZNN		All clear of traffic ("nothing now").
ZNR		Not received.
ZOH	What traffic have you on hand?	I have ______ groups on hand to transmit.
ZOK		We are receiving OK.
ZSF		Send faster.
ZSH		Static heavy here.
ZSR		Your signals strong and readable.
ZSS		Send slower.
ZSU		Your signals are unreadable.
ZSV		Your speed varying.
ZTA		Transmit by automatic.
ZTH		Transmit by hand.
ZVF		Your signal varying in frequency.
ZVP		Send V's please (for tuning).
ZWO		Send words once.
ZWT		Send words twice.

TABLE A-4
Factors for Converting Units

To convert A to D, multiply A by B. To convert D to A, multiply D by C.

Unit A	Factors B	C	Unit D
Length			
Miles	63,360.	0.00001578	Inches
Miles	5,280.	0.0001894	Feet
Miles	1.609	0.6214	Kilometers
Nautical miles	1.1508	0.869	Miles
Meters	3.281	0.3048	Feet
Kilometers	3,280.8	0.0003048	Feet
Inches	2.540	0.3937	Centimeters
Feet	.1667	6.0	Fathoms
Surface			
Square miles	27,878,400.	0.00000003537	Square feet
Square miles	640.	0.001563	Acres
Acres	43,560.	0.00002296	Square feet
Acres	4,047.	0.0002471	Square meters
Square inches	6.452	0.1550	Square centimeters
Square meters	10.76	0.09290	Square feet
Volume			
Cubic feet	0.025	40.0	Tons (shipping)
Cubic feet	1,728.	0.0005787	Cubic inches
Cubic inches	16.39	0.06102	Cubic centimeters
Cubic meters	35.31	0.02832	Cubic feet
Cubic feet	7.481	0.1337	U.S. gallons
Cubic feet	6.232	0.1605	Imperial gallons
Cubic feet	28.32	0.03531	Liters
U.S. gallons	231.	0.004329	Cubic inches
U.S. gallons	3.785	0.2042	Liters
Imperial gallons	1.201	0.8327	U.S. gallons
Fluid ounces	1.805	0.5540	Cubic inches
Velocities			
Miles per hour	1.467	0.6818	Feet per second
Meters per second	3.281	0.3048	Feet per second
Meters per second	2.237	0.4470	Miles per hour
Pressure			
Atmospheric (mean)	14.70	0.0680	Pounds per square inch
Atmospheric (mean)	29.92	0.03342	Inches of mercury
Pounds per square inch	2.036	0.4912	Inches of mercury
Feet of water	62.42	0.01602	Pounds per square foot

TABLE A-4
Factors for Converting Units

(continued)

To convert A to D, multiply A by B. To convert D to A, multiply D by C.

Unit A	Factors B	Factors C	Unit D
Weight			
Ounces	0.0625	16.0	Pounds
Pounds	7,000.0	0.0001429	Grains (avoir.)
Kilograms	2.205	0.4536	Pounds
Short tons	2,000.	0.0005	Pounds
Long tons	1.120	0.8929	Short tons
Angular measure			
Circle	360.0		Degrees
Degrees	60.0		Minutes
Degrees	17.8	0.056	Mils
Mils	3.37	0.297	Minutes
Minutes	60.		Seconds

TABLE A-5
Windchill Factor Chart

Estimated Wind Speed (In mph)	Actual Temperature Reading (°F)											
	50	40	30	20	10	0	-10	-20	-30	-40	-50	-60
	Equivalent Chill Temperature (°F)											
Calm	50	40	30	20	10	0	-10	-20	-30	-40	-50	-60
5	48	37	27	16	6	-5	-15	-26	-36	-47	-57	-68
10	40	28	16	4	-9	-24	-33	-46	-58	-70	-83	-95
15	36	22	9	-5	-18	-32	-45	-58	-72	-85	-99	-112
20	32	18	4	-10	-25	-39	-53	-67	-82	-96	-110	-121
25	30	16	0	-15	-29	-44	-59	-74	-88	-104	-118	-133
30	28	13	-2	-18	-33	-48	-63	-79	-94	-109	-125	-140
35	27	11	-4	-20	-35	-51	-67	-82	-98	-113	-129	-145
40	26	10	-6	-21	-37	-53	-69	-85	-100	-116	-132	-148

(Wind speeds greater then 40 mph have little additional effect.)

LITTLE DANGER
Is < hr with dry skin. Maximum danger of false sense of security.

INCREASING DANGER
Danger from freezing of exposed flesh within one minute.

GREAT DANGER
Flesh may freeze within 30 seconds

Trenchfoot and immersion foot may occur at any point on this chart.

Developed by U.S. Army Research Institute of Environmental Medicine, Natick, MA.

The windchill chart expresses the potential damage to exposed skin in a cold, windy environment.

TABLE A-6
Temperature—Humidity Index Table

Sum	THI	Sum	THI	Sum	THI	Sum	THI	Sum	THI
100	55	**120**	63	**140**	71	**160**	79	**180**	87
101	55	**121**	63	**141**	71	**161**	79	**181**	87
102	56	**122**	64	**142**	72	**162**	80	**182**	88
103	56	**123**	64	**143**	72	**163**	80	**183**	88
104	57	**124**	65	**144**	73	**164**	81	**184**	89
105	57	**125**	65	**145**	73	**165**	81	**185**	89
106	57	**126**	65	**146**	73	**166**	81	**186**	89
107	58	**127**	66	**147**	74	**167**	82	**187**	90
108	58	**128**	66	**148**	74	**168**	82	**188**	90
109	59	**129**	67	**149**	75	**169**	83	**189**	91
110	59	**130**	67	**150**	75	**170**	83	**190**	91
111	59	**131**	67	**151**	75	**171**	83	**191**	91
112	60	**132**	68	**152**	76	**172**	84	**192**	92
113	60	**133**	68	**153**	76	**173**	84	**193**	92
114	61	**134**	69	**154**	77	**174**	85	**194**	93
115	61	**135**	69	**155**	77	**175**	85	**195**	93
116	61	**136**	69	**156**	77	**176**	85	**196**	93
117	62	**137**	70	**157**	78	**177**	86	**197**	94
118	62	**138**	70	**158**	78	**178**	86	**198**	94
119	63	**139**	71	**159**	79	**179**	87	**199**	95

The Temperature-Humidity Index quantifies the degree of discomfort that results from high temperature and humidity. To derive the THI, add the sums of dry-bulb and wet-bulb temperatures taken simultaneously and locate the corresponding THI on the chart.

TABLE A-7 Time Zone Chart

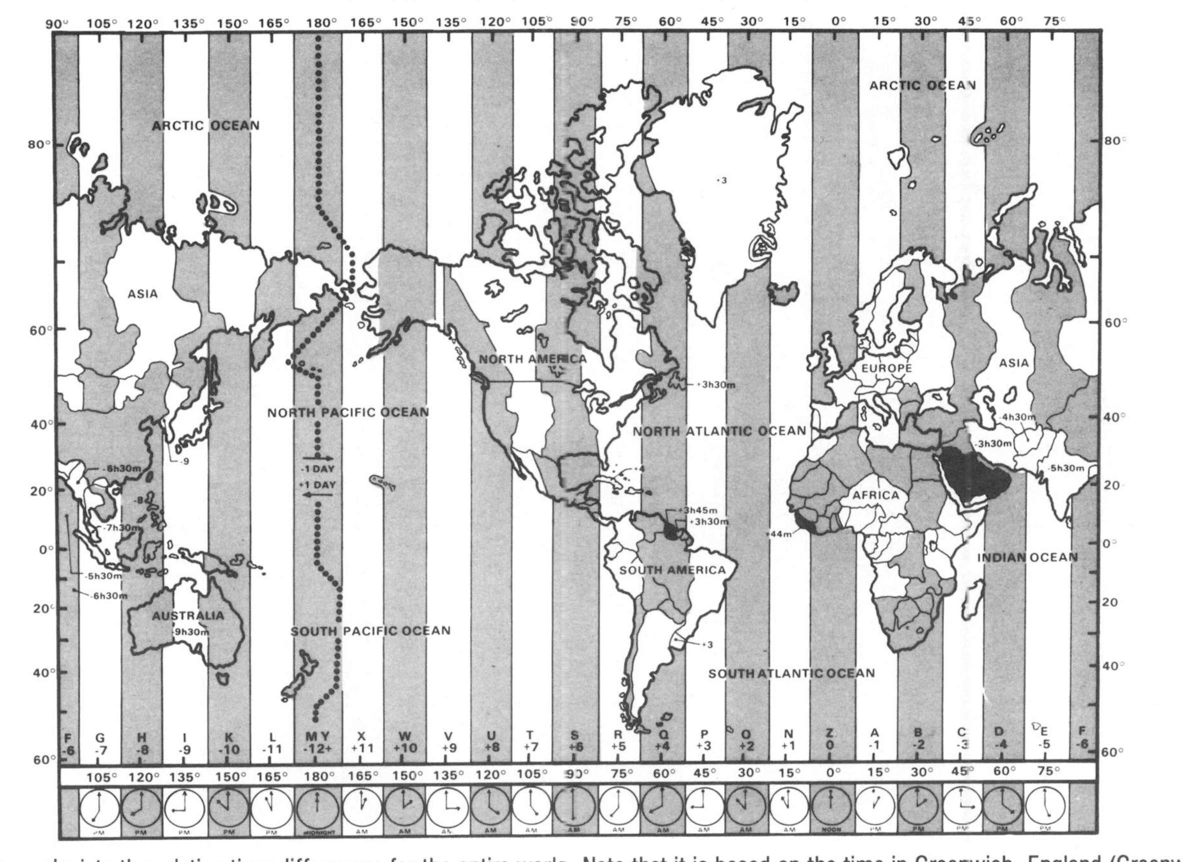

This map depicts the relative time differences for the entire world. Note that it is based on the time in Greenwich, England (Greenwich mean or zulu time) and is generally divided into 15-degree longitudinal segments per hour of difference. Local exceptions are common and subject to continuing modification by political authorities.

TABLE A-8
Julian Date Conversion Table (Perpetual)

Day	Jan	Feb	Mar	Apr	May	June	July	Aug	Sep	Oct	Nov	Dec	Day
1	001	032	060	091	121	152	182	213	244	274	305	335	**1**
2	002	033	061	092	122	153	183	214	245	275	306	336	**2**
3	003	034	062	093	123	154	184	215	246	276	307	337	**3**
4	004	035	063	094	124	155	185	216	247	277	308	338	**4**
5	005	036	064	095	125	156	186	217	248	278	309	339	**5**
6	006	037	065	096	126	157	187	218	249	279	310	340	**6**
7	007	038	066	097	127	158	188	219	250	280	311	341	**7**
8	008	039	067	098	128	159	189	220	251	281	312	342	**8**
9	009	040	068	099	129	160	190	221	252	282	313	343	**9**
10	010	041	069	100	130	161	191	222	253	283	314	344	**10**
11	011	042	070	101	131	162	192	223	254	284	315	345	**11**
12	012	043	071	102	132	163	193	224	255	285	316	346	**12**
13	013	044	072	103	133	164	194	225	256	286	317	347	**13**
14	014	045	073	104	134	165	195	226	257	287	318	348	**14**
15	015	046	074	105	135	166	196	227	258	288	319	349	**15**
16	016	047	075	106	136	167	197	228	259	289	320	350	**16**
17	017	048	076	107	137	168	198	229	260	290	321	351	**17**
18	018	049	077	108	138	169	199	230	261	291	322	352	**18**
19	019	050	078	109	139	170	200	231	262	292	323	353	**19**
20	020	051	079	110	140	171	201	232	263	293	324	354	**20**
21	021	052	080	111	141	172	202	233	264	294	325	355	**21**
22	022	053	081	112	142	173	203	234	265	295	326	356	**22**
23	023	054	082	113	143	174	204	235	266	296	327	357	**23**
24	024	055	083	114	144	175	205	236	267	297	328	358	**24**
25	025	056	084	115	145	176	206	237	268	298	329	359	**25**
26	026	057	085	116	146	177	207	238	269	299	330	360	**26**
27	027	058	086	117	147	178	208	239	270	300	331	361	**27**
28	028	059	087	118	148	179	209	240	271	301	332	362	**28**
29	029		088	119	149	180	210	241	272	302	333	363	**29**
30	030		089	120	150	181	211	242	273	303	334	364	**30**
31	031		090		151		212	243		304		365	**31**

TABLE A-9
Julian Date Conversion Table
(Leap Years)

Day	Jan	Feb	Mar	Apr	May	June	July	Aug	Sep	Oct	Nov	Dec	Day
1	001	032	061	092	122	153	183	214	245	275	306	336	1
2	002	033	062	093	123	154	184	215	246	276	307	337	2
3	003	034	063	094	124	155	185	216	247	277	308	338	3
4	004	035	064	095	125	156	186	217	248	278	309	339	4
5	005	036	065	096	126	157	187	218	249	279	310	340	5
6	006	037	066	097	127	158	188	219	250	280	311	341	6
7	007	038	067	098	128	159	189	220	251	281	312	342	7
8	008	039	068	099	129	160	190	221	252	282	313	343	8
9	009	040	069	100	130	161	191	222	253	283	314	344	9
10	010	041	070	101	131	162	192	223	254	284	315	345	10
11	011	042	071	102	132	163	193	224	255	285	316	346	11
12	012	043	072	103	133	164	194	225	256	286	317	347	12
13	013	044	073	104	134	165	195	226	257	287	318	348	13
14	014	045	074	105	135	166	196	227	258	288	319	349	14
15	015	046	075	106	136	167	197	228	259	289	320	350	15
16	016	047	076	107	137	168	198	229	260	290	321	351	16
17	017	048	077	108	138	169	199	230	261	291	322	352	17
18	018	049	078	109	139	170	200	231	262	292	323	353	18
19	019	050	079	110	140	171	201	232	263	293	324	354	19
20	020	051	080	111	141	172	202	233	264	294	325	355	20
21	021	052	081	112	142	173	203	234	265	295	326	356	21
22	022	053	082	113	143	174	204	235	266	296	327	357	22
23	023	054	083	114	144	175	205	236	267	297	328	358	23
24	024	055	084	115	145	176	206	237	268	298	329	359	24
25	025	056	085	116	146	177	207	238	269	299	330	360	25
26	026	057	086	117	147	178	208	239	270	300	331	361	26
27	027	058	087	118	148	179	209	240	271	301	332	362	27
28	028	059	088	119	149	180	210	241	272	302	333	363	28
29	029	060	089	120	150	181	211	242	273	303	334	364	29
30	030		090	121	151	182	212	243	274	304	335	365	30
31	031		091		152		213	244		305		366	31

TABLE A-10
Classes and Subclasses of Supply

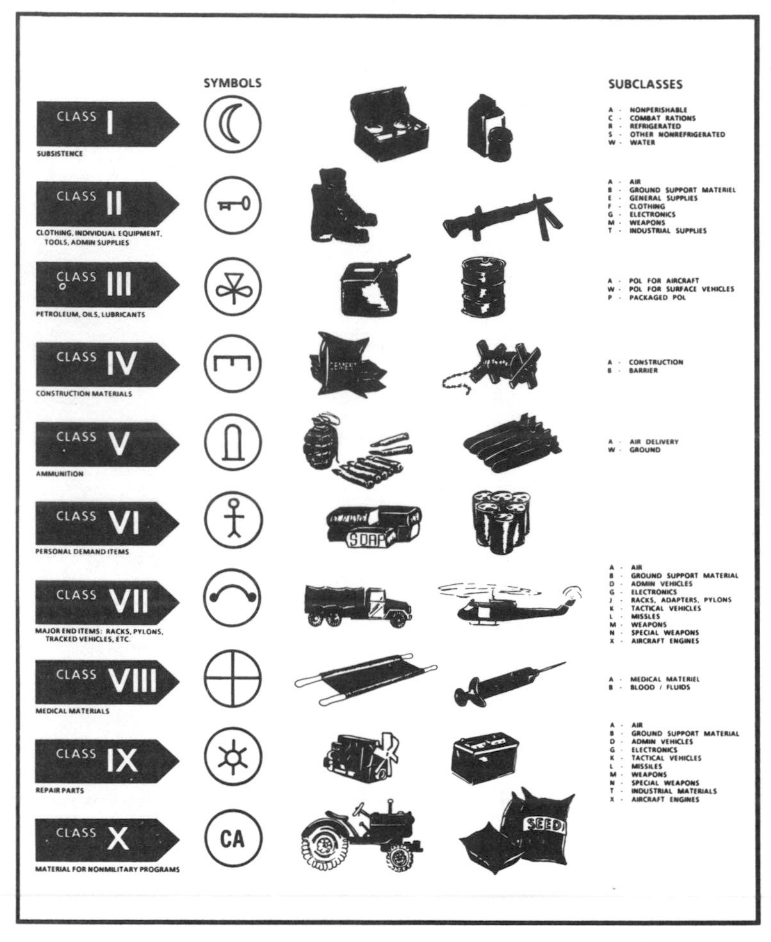

Source: FM 38-725. References: FM 101-5-1, *Operational Terms and Symbols;* JOPS III LCE Users Manual; Armed Forces Staff College Pub. 1.

TABLE A-11
Platoon Formations

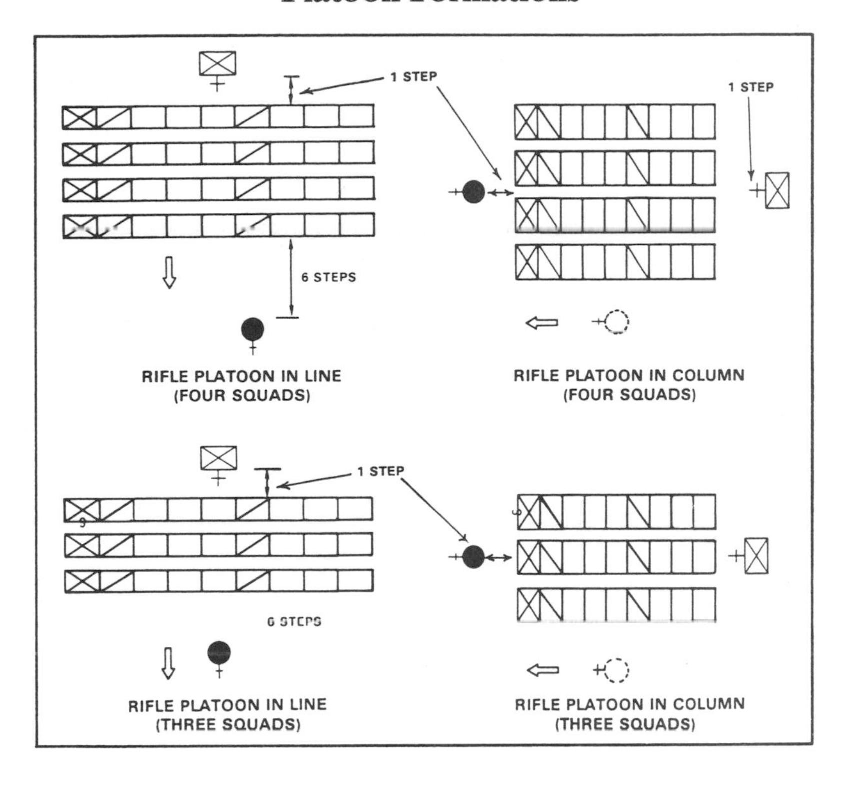

TABLE A-12
Company Formation

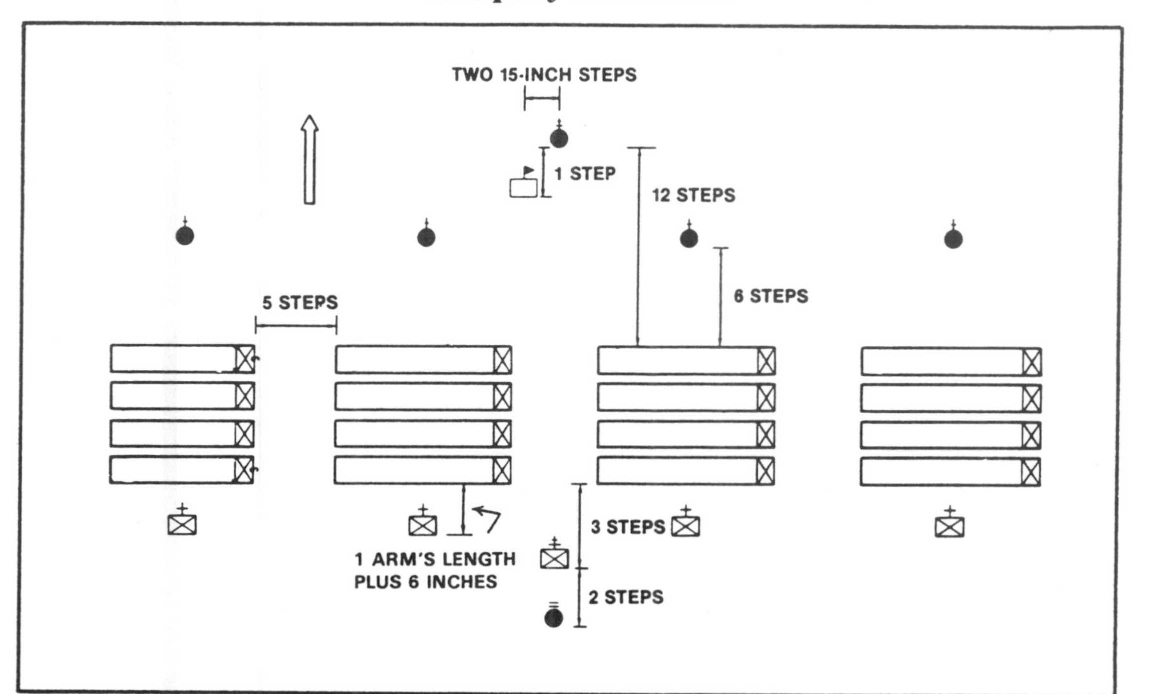

Company in line with platoons in line. Basic formations are depicted in IAW FM 22-5, which also describes higher-level formations.

TABLE A-13
Modern Army Recordkeeping System (MARKS)

MARKS replaced the Army Functional File System in 1986 as the standard for maintaining documents. These major categories are further broken down in AR 25-400-2.

1	Administration
5	Management
10	Organization and functions
11	Army programs
12	Security assistance
15	Boards, commissions, and committees
18	Army automation
20	Assistance, inspections, investigations, and follow-up
25	Information management
27	Legal services
30	Food program
32	Clothing and textile materiel
34	Standardization
36	Audit
37	Financial administration
40	Medical services
50	Nuclear and chemical weapons and materiel
55	Transportation and travel
56	Surface transportation
58	Motor transportation
59	Air transportation
60	Exchange service
65	Postal service
66	Courier service
70	Research, development, and acquisition
71	Force development
75	Explosives
95	Aviation
105	Communications-electronics
108	Audio-visual services
115	Climatic, hydrological, and topographic services
117	Corps of Engineers topography and geodesy
135	Army National Guard and Army Reserve
140	Army Reserve
145	Reserve Officers Training Corps
165	Religious activities
190	Military police
195	Criminal investigation
200	Environmental quality
210	Installations
215	Morale, Welfare, and Recreation
220	Field organizations
230	Nonappropriated funds and related activities
290	Cemeteries
310	Military publications
325	Statistics
335	Management information control
340	Office management
350	Training
351	Schools
352	Dependents' education
360	Army information
380	Security
381	Military intelligence
385	Safety
405	Real estate
415	Construction
420	Facilities engineering
500	Emergency employment of Army and other resources
525	Military operations
530	Operations and signal security
550	Foreign countries and nationals
570	Manpower and equipment control

TABLE A-13
Modern Army Recordkeeping System (MARKS)
(continued)

600	Personnel—general
601	Personnel procurement
602	Man-materiel systems
604	Personnel security
608	Personal affairs
611	Personnel selection and classification
612	Personnel processing
614	Assignments, details, and transfers
621	Education
623	Personnel evaluation
624	Promotions
630	Personnel absences
633	Apprehension and confinement
635	Personnel separations
638	Deceased personnel
640	Personnel records and identification of individuals
672	Decorations, awards, and honors
680	Personnel information systems
690	Civilian personnel
700	Logistics
702	Product assurance
703	Petroleum management
708	Cataloging of supplies and equipment
710	Inventory management
715	Procurement
725	Requisition and issue of supplies and equipment
735	Property accountability
738	Maintenance management
740	Storage and supply activities
746	Marking, packing, and shipment of supplies and equipment
750	Maintenance of supplies and equipment
755	Disposal of supplies and equipment
795	International logistics
840	Heraldic activities
870	Historic activities
920	Civilian marksmanship
930	Service organizations
1105	Corps of Engineers planning
1110	Corps of Engineers engineering and design
1125	Corps of Engineers plant
1130	Corps of Engineers project operation
1145	Corps of Engineers regulatory functions
1165	Corps of Engineers water resource policies and authority
1180	Corps of Engineers contracts

TABLE A-14
Numerical List of Enlisted Military Occupational Specialties

Code	Specialty
00B	Diver
00E	Recruiter (reserve components)
00R	Recruiter/retention NCO
00Z	Command sergeant major
01H	Biological sciences assistant
02B	Cornet or trumpet player
02C	Baritone or euphonium player
02D	French horn player
02E	Trombone player
02F	Tuba player
02G	Flute or piccolo player
02H	Oboe player
02J	Clarinet player
02K	Bassoon player
02L	Saxophone player
02M	Percussion player
02N	Piano player
02P	Brass group leader
02Q	Woodwind group leader
02S	Special bands member
02T	Guitar player
02U	Electric bass player
02Z	Bands senior sergeant
11B	Infantryman
11C	Indirect fire infantryman
11H	Heavy antiarmor weapons infantryman
11M	Fighting vehicle infantryman
11Z	Infantry senior sergeant
12B	Combat engineer
12C	Bridge crewmember
12F	Engineer tracked vehicle crewman
12Z	Combat engineering senior sergeant
13B	Cannon crewmember
13C	TACFIRE operations specialist
13E	Cannon fire direction specialist
13F	Fire support specialist
13M	Multiple launch rocket system crewmember
13N	Lance crewmember
13P	Multiple launch rocket system/ Lance operations/fire direction specialist
13R	Field artillery firefinder radar operator
13T	Remotely piloted vehicle crewmember
13Z	Field artillery senior sergeant
16D	Hawk missile crewmember
16E	Hawk fire control crewmember
16F	Light air defense artillery crewmember (reserve components)
16J	Forward area alerting operator
16P	Chapparal crewmember
16R	Vulcan crewmember
16S	Man-portable air defense or pedestal-mounted STINGER crewmember
16T	PATRIOT missile crewmember
16Z	Air defense artillery senior sergeant
17B	Field artillery radar crewmember (reserve components)
18B	Special forces weapons sergeant
18C	Special forces engineer sergeant
18D	Special forces medical sergeant
18E	Special forces communications sergeant
18F	Special forces assistant operations and intelligence sergeant
18Z	Special forces senior sergeant
19D	Cavalry scout
19E	M48-M60 armor crewman
19K	M1 armor crewman
19Z	Armor senior sergeant
21G	Pershing electronics materiel specialist
21L	Pershing electronics repairer
23R	Hawk missile system mechanic
24C	Hawk firing section mechanic

TABLE A-14
Numerical List of Enlisted Military Occupational Specialties
(continued)

24G	Hawk information coordination central mechanic
24H	Hawk fire control repairer
24K	Hawk continuous wave radar repairer
24M	Vulcan system mechanic
24N	Chapparal system mechanic
24R	Hawk master mechanic
24T	PATRIOT operator and system mechanic
25L	AN/TSQ 73 air defense artillery command and control system operator/repairer
25P	Visual information/audio documentation systems specialist
25Q	Graphics documentation technician
25R	Visual information/audio equipment repairer
25S	Still documentation specialist
25Z	Visual information chief
27B	Land combat support system test specialist
27E	TOW/Dragon repairer
27F	Vulcan repairer
27G	Chapparal/Redeye repairer
27H	HAWK firing section repairer
27J	HAWK field maintenance equipment/pulse acquisition radar repairer
27K	Hawk fire control/continuous wave radar repairer
27L	Lance system repairer
27M	Multiple-launch rocket system repairer
27N	Forward area alerting radar repairer
27T	Pedestal-mounted STINGER, line-of-sight rear air defense system repairer
27V	Hawk maintenance chief
27X	PATRIOT system repairer
27Z	Land combat/air defense systems maintenance chief
29E	Radio repairer
29F	Fixed communications security equipment repairer
29J	Telecommunications terminal device repairer
29M	Tactical satellite/microwave repairer
29N	Telephone central office repairer
29P	Communications security maintenance chief
29S	Field communications security equipment repairer
29T	Satellite/microwave communications chief
29V	Strategic microwave systems repairer
29W	Communications maintenance support chief
29X	Communications equipment maintenance chief
29Y	Satellite communications systems repairer
29Z	Electronics maintenance chief
31C	Single channel radio operator
31D	Mobile subscriber equipment transmission system operator
31F	Mobile subscriber equipment network switching system operator
31G	Tactical communications chief
31K	Combat signaller
31L	Wire systems installer
31M	Multichannel communications systems operator
31N	Communications systems/circuit controller
31Q	Tactical satellite/microwave systems operator

TABLE A-14
Numerical List of Enlisted Military Occupational Specialties
(continued)

31V	Unit-level communications maintainer
31W	Mobile subscriber equipment communications chief
31Y	Communications systems supervisor
31Z	Communications operations chief
33M	Electronic warfare/intercept strategic systems analyst and command and control subsystems repairer
33P	Electronic warfare/intercept strategic receiving subsystems repairer
33Q	Electronic warfare/intercept strategic processing and storage subsystems repairer
33R	Electronic warfare/intercept aviation systems repairer
33T	Electronic warfare/intercept tactical systems repairer
33V	Electronic warfare/intercept aerial sensor repairer
33Z	Electronic warfare/intercept systems maintenance supervisor
35G	Medical equipment repairer, unit level
35H	Test, measurement, and diagnostic equipment maintenance support specialist
35U	Medical equipment repairer, advanced
36L	Transportable automatic switching systems operator/maintainer
36M	Switching systems operator
39B	Automatic test equipment operator/maintainer
39C	Target acquisition/surveillance radar repairer
39D	Decentralized automated service support system computer systems repairer
39E	Special electronic devices repairer
39G	Automated communications computer systems repairer
39L	Field artillery digital systems repairer
39V	Computerized systems maintenance chief
39W	Radar/special electronic devices maintenance chief
39X	Electronics equipment maintenance chief
39Y	Field artillery tactical fire direction systems repairer
41C	Fire control instrument repairer
42C	Orthotic specialist
42D	Dental laboratory specialist
42E	Optical laboratory specialist
43E	Parachute rigger
43M	Fabric repair specialist
44B	Metal worker
44E	Machinist
45B	Small arms repairer
45D	Self-propelled field artillery turret mechanic
45E	M1 Abrams tank turret mechanic
45G	Fire control system repairer
45K	Tank turret repairer
45L	Artillery repairer
45N	M60A1/A3 tank turret mechanic
45T	Bradley fighting vehicle system turret mechanic
45Z	Armament/fire control maintenance supervisor
46N	Pershing electrical-mechanical repairer
46Q	Journalist
46R	Broadcast journalist
46Z	Public affairs chief
51B	Carpentry and masonry specialist
51G	Materials quality specialist

TABLE A-14
Numerical List of Enlisted Military Occupational Specialties
(continued)

51H	Construction engineering supervisor
51K	Plumber
51M	Firefighter
51R	Interior electrician
51T	Technical engineering supervisor
51Z	General engineering supervisor
52C	Utilities equipment repairer
52D	Power generation equipment repairer
52E	Prime power production specialist
52F	Turbine engine driven generator repairer
52G	Transmission and distribution specialist (reserve components)
52X	Special purpose equipment repairer
54B	Chemical operations specialist
55B	Ammunition specialist
55D	Explosive ordnance disposal specialist
55G	Nuclear weapons specialist
55R	Ammunition stock control and accounting specialist
55X	Ammunition inspector
55Z	Ammunition supervisor
57E	Laundry and bath specialist
57F	Graves registration specialist
62B	Construction equipment repairer
62E	Heavy construction equipment operator
62F	Crane operator
62G	Quarrying specialist
62H	Concrete and asphalt equipment operator
62J	General construction equipment operator
62N	Construction equipment supervisor
63B	Light-wheel vehicle mechanic
63D	Self-propelled field artillery system mechanic
63E	M1 Abrams tank system mechanic
63G	Fuel and electrical systems repairer
63H	Track vehicle repairer
63J	Quartermaster and chemical equipment repairer
63N	M60A1/A3 tank system mechanic
63S	Heavy-wheel vehicle mechanic
63T	Bradley fighting vehicle system mechanic
63W	Wheel vehicle repairer
63Y	Track vehicle mechanic
63Z	Mechanical maintenance supervisor
67G	Utility airplane repairer (reserve components)
67H	Observation airplane repairer
67N	UH-1 helicopter repairer
67R	AH-64 attack helicopter repairer
67S	OH-58D helicopter repairer
67T	UH-60 helicopter repairer
67U	CH-47 helicopter repairer
67V	Observation/scout helicopter repairer
67X	Heavy lift helicopter repairer (reserve components)
67Y	AH-1 attack helicopter repairer
67Z	Aircraft maintenance senior sergeant
68B	Aircraft power plant repairer
68D	Aircraft power train repairer
68F	Aircraft electrician
68G	Aircraft structural repairer
68H	Aircraft pneudraulic repairer
68J	Aircraft armament/missile systems repairer
68K	Aircraft components repair supervisor

TABLE A-14
Numerical List of Enlisted Military Occupational Specialties
(continued)

68L	Avionic communications equipment repairer
68N	Avionic mechanic
68P	Avionic maintenance supervisor
68Q	Avionic flight systems repairer
68R	Avionic radar repairer
71C	Executive administrative assistant
71D	Legal specialist
71E	Court reporter
71G	Patient administration specialist
71L	Administrative specialist
71M	Chaplain assistant
72E	Tactical telecommunications center operator
72G	Automatic data telecommunications center operator
73C	Finance specialist
73D	Accounting specialist
73Z	Finance senior sergeant
74D	Computer/machine operator
74F	Programmer/analyst
74Z	Data processing NCO
75B	Personnel administration specialist
75C	Personnel management specialist
75D	Personnel records specialist
75E	Personnel action specialist
75F	Personnel information system management specialist
75Z	Personnel sergeant
76C	Equipment records and parts specialist
76J	Medical supply specialist
76P	Materiel control and accounting specialist
76V	Materiel storage and handling specialist
76X	Subsistence supply specialist
76Y	Unit supply specialist
76Z	Senior supply/service sergeant
77F	Petroleum supply specialist
77L	Petroleum laboratory specialist
77W	Water treatment specialist
79D	Reenlistment NCO (reserve component)
81B	Technical drafting specialist
81C	Cartographer
81Q	Terrain analyst
81Z	Topographic engineering supervisor
82B	Construction surveyor
82C	Field artillery surveyor
82D	Topographic surveyor
83E	Photo and layout specialist
83F	Printing and bindery specialist
88H	Cargo specialist
88K	Watercraft operator
88L	Watercraft engineer
88M	Motor transport operator
88N	Traffic management coordinator
88P	Locomotive repairer (reserve components)
88Q	Railway car repairer (reserve components)
88R	Airbrake repairer (reserve components)
88S	Locomotive electrician (reserve components)
88T	Railway section repairer (reserve components)
88U	Locomotive operator (reserve components)
88V	Train crewmember (reserve components)
88W	Railway movement coordinator (reserve components)
88X	Railway senior sergeant (reserve components)

TABLE A-14
Numerical List of Enlisted Military Occupational Specialties
(continued)

88Y	Marine senior sergeant
88Z	Transportation senior sergeant
91A	Medical specialist
91B	Medical NCO
91C	Practical nurse
91D	Operating room specialist
91E	Dental specialist
91F	Psychiatric specialist
91G	Behavioral science specialist
91H	Orthopedic specialist
91J	Physical therapy specialist
91L	Occupational therapy specialist
91N	Cardiac specialist
91P	X-ray specialist
91Q	Pharmacy specialist
91R	Veterinary food inspection specialist
91S	Preventive medicine specialist
91T	Animal care specialist
91U	Ear, nose, and throat specialist
91V	Respiratory specialist
91W	Nuclear medicine specialist
91X	Health physics specialist
91Y	Eye specialist
92B	Medical laboratory specialist
92E	Cytology specialist
93B	Aeroscout observer
311A	CID special agent
350B	All-source intelligence technician
350D	Imagery intelligence technician
350L	Attaché technician
351B	Counterintelligence technician
351C	Area intelligence technician
351E	Interrogation technician
352C	Traffic analysis technician
352D	Emitter location/identification technician
352G	Voice intercept technician
352H	Morse intercept technician
352J	Emanations intercept technician
352K	Non-Morse intercept technician
353A	IEW equipment technician
420A	Military personnel technician
420C	Bandmaster
420D	Club manager
550A	Legal administrator
600A	Physician assistant
640A	Veterinary services technician
670A	Veterinary services maintenance technician
880A	Marine deck officer
881A	Marine engineering officer
910A	Ammunition technician
911A	Nuclear weapons technician
912A	Land combat missile systems technician
913A	Armament repair technician
914A	Allied trades technician
915A	Wheel vehicle maintenance technician
915B	Light-track systems maintenance technician
915C	Field artillery systems maintenance technician
915D	Armor/cavalry systems maintenance technician
915E	Support/staff maintenance technician
920A	Property accounting technician
920B	Supply systems technician
921A	Airdrop systems technician
922A	Food service technician

For details on enlisted specialties see AR 611-201.

TABLE A-15
Numerical List of Officer Specialties

01A	Branch immaterial
02A	Combat arms immaterial
03A	Logistics immaterial
04A	Personnel immaterial
11A	Infantry general
11B	Light infantry
11C	Mechanized infantry
12A	Armor, general
12B	Armor
12C	Cavalry
13A	Field artillery, general
13B	Light missile field artillery
13C	Heavy missile field artillery
13D	Field artillery target acquisition
13E	Cannon field artillery
14A	Air defense artillery, general
14B	Short-range air defense artillery (SHORAD)
14D	Hawk missile ADA
14E	PATRIOT missile ADA
15A	Aviation, general
15B	Aviation, combined arms operations
15C	Aviation, tactical intelligence
15D	Aviation, logistics
15E	Aviation, tactical communications (air traffic control)
18A	Special forces
21A	Engineer, general
21B	Combat engineer
21C	Topographic engineer
21D	Facilities/contract construction management engineer (FCCME)
25A	Signal, general
25B	Communications-electronics (C-E) automation
25C	Communications-electronics (C-E) operations
25D	Communications-electronics (C-E) engineering
25E	Information systems and networking
31A	Military police, general
31B	Physical security
31C	Correctional
31D	Criminal investigation
35A	Military intelligence, general
35B	Strategic intelligence
35C	Imagery intelligence
35D	Tactical intelligence
35E	Counterintelligence
35F	Human intelligence
35G	Signals intelligence/electronic warfare
38A	Civil affairs, general
42A	Adjutant general, general
42B	Personnel systems management
42C	Army band
42E	Administrative systems management
44A	Finance, general
55A	Judge advocate
55B	Military judge
56A	Command and unit chaplain
56D	Clinical pastoral educator
60A	Operational medicine
60B	Nuclear medicine officer
60C	Preventive medicine officer
60D	Occupational medicine officer
60F	Pulmonary disease officer
60G	Gastroenterologist
60H	Cardiologist
60J	Obstetrician and gynecologist
60K	Urologist
60L	Dermatologist
60M	Allergist, clinical immunologist
60N	Anesthesiologist
60P	Pediatrician
60Q	Pediatric cardiologist
60R	Child neurologist
60S	Ophthalmologist
60T	Otolaryngologist
60U	Child psychiatrist

TABLE A-15
Numerical List of Officer Specialties
(continued)

Code	Specialty
60V	Neurologist
60W	Psychiatrist
61A	Nephrologist
61B	Medical oncologist/hematologist
61C	Endocrinologist
61D	Rheumatologist
61E	Clinical pharmacologist
61F	Internist
61G	Infectious disease officer
61H	Family physician
61J	General surgeon
61K	Thoracic surgeon
61L	Plastic surgeon
61M	Orthopedic surgeon
61N	Flight surgeon
61P	Physiatrist
61Q	Therapeutic radiologist
61R	Diagnostic radiologist
61U	Pathologist
61W	Peripheral vascular surgeon
61Z	Neurosurgeon
62A	Emergency physician
62B	Field surgeon
63A	Dentistry, general
63B	Dentistry, comprehensive
63D	Periodontist
63E	Endodontist
63F	Prosthodontics
63H	Preventive dentistry/dental public health
63K	Pedodontist
63M	Orthodontist
63N	Oral surgeon
63P	Oral pathologist
63R	Executive dental officer
64A	Veterinary services officer
64B	Veterinary staff officer
64C	Veterinary laboratory animal medicine officer
64D	Veterinary pathologist
64E	Veterinary microbiologist
64F	Veterinary comparative medicine officer
65A	Occupational therapy
65B	Physical therapy
65C	Hospital dietician
66A	Nurse administrator
66B	Community health nurse
66C	Psychiatric/mental health nurse
66D	Pediatric nurse
66E	Operating room nurse
66F	Nurse anesthetist
66G	Obstetric and gynecologic nurse
66H	Medical-surgical nurse
66J	Clinical nurse
67A	Health care administration
67B	Field medical assistant
67C	Health services comptroller
67D	Biomedical information systems officer
67E	Patient administration officer
67F	Health services personnel manager
67G	Health services manpower control officer
67H	Health services plans, operations, intelligence, and training officer
67J	Aeromedical evacuation officer
67K	Health services materiel officer
67L	Health facilities planning officer
68A	Microbiologist
68B	Nuclear medical science officer
68C	Biochemist
68D	Parasitologist
68E	Immunologist
68F	Clinical laboratory officer/laboratory manager
68G	Entomologist
68H	Pharmacy officer
68J	Physiologist

TABLE A-15
Numerical List of Officer Specialties
(continued)

68K	Optometry officer
68L	Podiatrist
68M	Audiologist
68N	Environmental science officer
68P	Sanitary engineer
68R	Social work officer
68S	Clinical psychologist
68T	Research psychologist
68U	Psychology associate
74A	Chemical, general
74B	Chemical operations and training
74C	Chemical munitions and materiel management
88A	Transportation, general
88B	Chemical operations and training
88C	Marine and terminal operations
88D	Motor/rail transportation
88E	Transportation management
91A	Ordnance, general
91B	Tank/automotive materiel management
91C	Missile materiel management
91D	Munitions materiel management
91E	Explosive ordnance disposal
92A	Quartermaster, general
92B	Supply and materiel management
92D	Aerial delivery and materiel
92F	Petroleum
92G	Subsistence
39A	Psychological operations or civil affairs
39B	Psychological operations
39C	Civil affairs
41A	Personnel programs management staff
45A	comptroller
45B	Program/budget
46A	Public affairs/general
46B	Broadcast
47A	Permanent professor
47B	Permanent associate professor
48A	Foreign area, general
48B	Latin America
48C	West Europe
48D	South Asia
48E	Russia/East Europe
48F	China
48G	Middle East/North Africa
48H	Northeast Asia
48I	Southeast Asia
48J	Africa, south of the Sahara
49A	Operations research, general
49B	Operations research, personnel
49C	Operations research, combat operations/materiel systems
49D	Operations research, planning, programming, and resource management
49E	Operations research, test and evaluation
49W	Trained, operations research/ systems analysis
49X	Untrained, operations research/ systems analysis
50A	Force development
51A	Research and development, general
51B	Test and evaluation
51C	Combat developments
51D	Acquisition
52A	Nuclear weapons, general
52B	Nuclear weapons research
53A	Software engineering officer
53B	Hardware engineering officer
53C	Automation management officer
54A	Operations, plans and training
97A	Contracting and industrial management officer

For details on officer specialties see AR 611-101.

TABLE A-16
APFT Scoring Standards—Push-ups

Scoring standards are used for converting raw scores to point scores after test events are completed

▼ INDICATES FEMALE STANDARDS

Repetitions \ AGE GROUP	17-21	17-21 ▼	22-26	22-26 ▼	27-31	27-31 ▼	32-36	32-36 ▼	37-41	37-41 ▼	42-46	42-46 ▼	47-51	47-51 ▼	52 +	52 + ▼
82	100 POINTS															
81	99															
80	98		100													
79	97		99													
78	96		98		100											
77	95		97		99											
76	94		96		98											
75	93		95		97											
74	92		94		96											
73	91		93		95		100									
72	90		92		94		99		100							
71	89		91		93		98		99							
70	88		90		92		97		98							
69	87		89		91		96		97							
68	86		88		90		95		96							
67	85		87		89		94		95							
66	84		86		88		93		94		100					
65	83		85		87		92		93		99					
64	82		84		86		91		92		98					
63	81		83		85		90		91		97					
62	80		82		84		89		90		96		100			
61	79		81		83		88		89		95		99			
60	78		80		82		87		88		94		98			
59	77		79		81		86		87		93		97			
58	76	100	78		80		85		86		92		96			
57	75	99	77		79		84		85		91		95			
56	74	98	76	100	78		83		84		90		94		100	
55	73	97	75	99	77		82		83		89		93		99	
54	72	96	74	98	76	100	81		82		88		92		98	
53	71	95	73	97	75	99	80		81		87		91		97	
52	70	94	72	96	74	98	79	100	80		86		90		96	
51	69	93	71	95	73	97	78	99	79		85		89		95	
50	68	92	70	94	72	96	77	98	78		84		88		94	
49	67	91	69	93	71	95	76	97	77		83		87		93	
48	66	90	68	92	70	94	75	96	76	100	82		86		92	
47	65	89	67	91	69	93	74	95	75	99	81		85		91	
46	64	88	66	90	68	92	73	94	74	98	80		84		90	
45	63	87	65	89	67	91	72	93	73	97	79	100	83		89	
44	62	86	64	88	66	90	71	92	72	96	78	99	82		88	
43	61	85	63	87	65	89	70	91	71	95	77	98	81		87	
42	60	84	62	86	64	88	69	90	70	94	76	97	80		86	
41	59	83	61	85	63	87	68	89	69	93	75	96	79	100	85	
40	58	82	60	84	62	86	67	88	68	92	74	95	78	99	84	100
39	57	81	59	83	61	85	66	87	67	91	73	94	77	98	83	99
38	56	80	58	82	60	84	65	86	66	90	72	93	76	97	82	98
37	55	79	57	81	59	83	64	85	65	89	71	92	75	96	81	97
36	54	78	56	80	58	82	63	84	64	88	70	91	74	95	80	96
35	53	77	55	79	57	81	62	83	63	87	69	90	73	94	79	95
34	52	76	54	78	56	80	61	82	62	86	68	89	72	93	78	94
33	51	75	53	77	55	79	60	81	61	85	67	88	71	92	77	93
32	50	74	52	76	54	78	59	80	60	84	66	87	70	91	76	92
31	49	73	51	75	53	77	58	79	59	83	65	86	69	90	75	91
30	48	72	50	74	52	76	57	78	58	82	64	85	68	89	74	90
29	47	71	49	73	51	75	56	77	57	81	63	84	67	88	73	89
28	46	70	48	72	50	74	55	76	56	80	62	83	66	87	72	88
27	45	69	47	71	49	73	54	75	55	79	61	82	65	86	71	87
26	44	68	46	70	48	72	53	74	54	78	60	81	64	85	70	86
25	43	67	45	69	47	71	52	73	53	77	59	80	63	84	69	85
24	42	66	44	68	46	70	51	72	52	76	58	79	62	83	68	84
23	41	65	43	67	45	69	50	71	51	75	57	78	61	82	67	83
22	40	64	42	66	44	68	48	70	50	74	56	77	60	81	66	82
21	39	63	41	65	42	67	46	69	48	73	55	76	58	80	65	81
20	38	62	40	64	40	66	44	68	46	72	54	75	56	79	64	80
19	37	61	38	63	38	65	42	67	44	71	52	74	54	78	63	79
18	36	60	36	63	36	64	40	66	42	70	50	72	52	77	62	78
17	34	58	34	61	34	63	38	65	40	68	48	70	50	76	61	77
16	32	56	32	60	32	62	36	64	38	66	46	68	48	75	60	76
15	30	54	30	58	30	60	34	62	36	64	44	68	46	74	57	75
14	28	52	28	56	28	58	32	60	34	62	42	64	44	72	54	74
13	26	50	26	54	26	56	30	58	32	60	39	62	43	70	51	72
12	24	48	24	52	24	54	28	56	30	58	36	60	42	68	48	70
11	22	44	22	50	22	52	26	54	28	56	33	58	38	64	44	68
10	20	40	20	46	20	50	24	52	26	54	30	56	36	60	40	64
9	18	36	18	42	18	45	22	50	24	52	27	54	34	57	36	60
8	16	32	16	38	16	40	20	45	22	50	24	52	32	54	32	56
7	14	28	14	34	14	35	18	40	20	44	21	50	28	51	28	52
6	12	24	12	30	12	30	16	35	18	38	18	43	24	48	24	48
5	10	20	10	25	10	25	14	30	15	32	15	36	20	40	20	40
4	8	16	8	20	8	20	12	24	12	26	12	29	16	32	16	32
3	6	12	6	15	6	15	9	18	9	20	9	22	12	24	12	24
2	4	8	4	10	4	10	6	12	6	14	6	15	8	16	8	16
1	2	4	2	5	2	5	3	6	3	7	3	8	4	8	4	8

TABLE A-17
APFT Scoring Standards—Sit-ups

Repetitions	AGE GROUP 17–21 (POINTS)	▼	22–26	▼	27–31	▼	32–36	▼	37–41	▼	42–46	▼	47–51	▼	52 +	▼
92	100															
91	99															
90	98	100														
89	97	99														
88	96	98														
87	95	97	100													
86	94	95	99													
85	93	94	98	100												
84	92	94	97	99												
83	91	93	96	98												
82	90	92	95	97	100											
81	89	91	94	96	99											
80	88	90	93	95	98	100										
79	87	89	92	94	97	99										
78	86	88	91	93	96	98	100									
77	85	87	90	92	95	97	99									
76	84	86	89	91	94	96	98									
75	83	85	88	90	93	95	97	100								
74	82	84	87	89	92	94	96	99								
73	81	83	86	88	91	93	95	98	100							
72	80	82	85	87	90	92	94	97	99							
71	79	81	84	86	89	91	93	96	98							
70	78	80	83	85	88	90	92	95	97	100						
69	77	79	82	84	87	89	91	94	96	99	100					
68	76	78	81	83	86	88	90	93	95	98	99					
67	75	77	80	82	85	87	89	92	94	97	98	100	100			
66	74	76	79	81	84	86	88	91	93	96	97	99	99		100	
65	73	75	78	80	83	85	87	90	92	95	96	98	98		99	
64	72	74	77	79	82	84	86	89	91	94	95	97	97	100	98	
63	71	73	76	78	81	83	85	88	90	93	94	96	96	99	97	
62	70	72	75	77	80	82	84	87	89	92	93	95	95	98	96	100
61	69	71	74	76	79	81	83	86	88	91	92	94	94	97	95	99
60	68	70	73	75	78	80	82	85	87	90	91	93	93	96	94	98
59	67	69	72	74	77	79	81	84	86	89	90	92	92	95	93	97
58	66	68	71	73	76	78	80	83	85	88	89	91	91	94	92	96
57	65	67	70	72	75	77	79	82	84	87	88	90	90	93	91	95
56	64	66	69	71	74	76	78	81	83	86	87	89	89	92	90	94
55	63	65	68	70	73	75	77	80	82	85	86	88	88	91	89	93
54	62	64	67	69	72	74	76	79	81	84	85	87	87	90	88	92
53	61	63	66	68	71	73	75	78	80	83	84	86	86	89	87	91
52	60	62	65	67	70	72	74	77	79	82	83	85	85	88	86	90
51	[illegible]	[illegible]	[illegible]	[illegible]	[illegible]	71	[illegible]	[illegible]	[illegible]	[illegible]	[illegible]	[illegible]	[illegible]	[illegible]	[illegible]	[illegible]
50	58	60	63	65	68	70	72	75	77	80	81	83	83	86	84	88
49	57	59	62	64	67	69	71	74	76	79	80	82	82	85	83	87
48	56	58	61	63	66	68	70	73	75	78	79	81	81	84	82	86
47	55	57	60	62	65	67	69	72	74	77	78	80	80	83	81	85
46	54	56	59	61	64	66	68	71	73	76	77	79	79	82	80	84
45	53	55	58	60	63	65	67	70	72	75	76	78	78	81	79	83
44	52	54	57	59	62	64	66	69	71	74	75	77	77	80	78	82
43	51	53	56	58	61	63	65	68	70	73	74	76	76	79	77	81
42	50	52	55	57	60	62	64	67	69	72	73	75	75	78	76	80
41	49	51	54	56	59	61	63	66	68	71	72	74	74	77	75	79
40	48	50	53	55	58	60	62	65	67	70	71	73	73	76	74	78
39	47	49	52	54	57	59	61	64	66	69	70	72	72	75	73	77
38	46	48	51	53	56	58	60	63	65	68	69	71	71	74	72	76
37	45	47	50	52	55	57	59	62	64	67	68	70	70	73	71	75
36	44	46	49	51	54	56	58	61	63	66	67	69	69	72	70	74
35	43	45	48	50	53	55	57	60	62	65	66	68	68	71	69	73
34	42	44	47	49	52	54	56	59	61	64	65	67	67	70	68	72
33	41	43	46	48	51	53	55	58	60	63	64	66	66	69	67	71
32	40	42	45	47	50	52	54	57	59	62	63	65	65	68	66	70
31	39	41	44	46	49	51	53	56	58	61	62	64	64	67	65	69
30	38	40	43	45	48	50	52	55	57	60	61	63	63	66	64	68
29	37	39	42	44	47	49	51	54	56	58	60	62	62	65	63	67
28	36	38	41	43	46	48	50	53	55	56	58	61	61	64	62	66
27	35	37	40	42	45	47	49	52	54	54	56	60	60	63	61	65
26	34	36	39	41	44	46	48	51	52	52	54	58	58	62	60	64
25	33	35	38	40	43	45	47	50	50	50	52	56	56	61	58	63
24	32	34	37	39	42	44	46	48	48	48	50	54	54	60	56	62
23	31	33	36	38	41	43	45	46	46	46	48	52	52	58	54	61
22	30	32	35	37	40	42	44	44	44	44	46	50	50	56	52	60
21	29	31	34	36	39	41	42	42	42	42	44	48	48	54	50	58
20	28	30	33	35	38	40	40	40	40	40	42	46	46	52	48	56
19	27	29	32	34	37	38	38	38	38	38	40	44	44	50	46	54
18	26	28	31	32	36	36	36	36	36	36	38	42	42	48	44	52
17	25	27	30	31	34	34	34	34	34	34	36	40	40	46	42	50
16	24	26	29	30	32	32	32	32	32	32	34	38	38	44	40	48
15	23	25	28	29	30	30	30	30	30	30	32	36	36	42	38	45
14	22	24	27	28	28	28	28	28	28	28	30	34	34	40	36	42
13	21	23	26	26	26	26	26	26	26	26	28	32	32	38	34	39
12	20	22	24	24	24	24	24	24	24	24	26	30	30	36	32	36
11	19	21	22	22	22	22	22	22	22	22	24	28	28	33	30	33
10	18	20	20	20	20	20	20	20	20	20	22	26	26	30	28	30
9	17	18	18	18	18	18	18	18	18	18	20	24	24	27	26	27
8	16	16	16	16	16	16	16	16	16	16	18	22	22	24	24	24
7	14	14	14	14	14	14	14	14	14	14	16	20	20	21	21	21
6	12	12	12	12	12	12	12	12	12	12	14	18	18	18	18	18
5	10	10	10	10	10	10	10	10	10	10	12	15	15	15	15	15
4	8	8	8	8	8	8	8	8	8	8	10	12	12	12	12	12
3	6	6	6	6	6	6	6	6	6	6	8	8	9	9	8	8
2	4	4	4	4	4	4	4	4	4	4	6	6	6	6	6	6
1	2	2	2	2	2	2	2	2	2	2	3	3	3	3	3	3

▼ INDICATES FEMALE STANDARDS

TABLE A-18
APFT Scoring Standards—Two-mile Run

▼ INDICATES FEMALE STANDARDS

Time / AGE GROUP	17–21	17–21 ▼	22–26	22–26 ▼	27–31	27–31 ▼	32–36	32–36 ▼	37–41	42–46	47–51	52 +
11:54	100											
12:00	99											
12:06	98											
12:12	97											
12:18	96											
12:24	95											
12:30	94											
12:36	93		100									
12:42	92		99									
12:48	91		98									
12:54	90		97									
13:00	89		96									
13:06	88		95									
13:12	87		94									
13:18	86		93		100							
13:20	85		92		99							
13:24	84		91		98							
13:36	83		90		97							
13:42	82		89		96							
13:48	81		88		95							
13:54	80		87		94							
14:00	79		86		93		100					
14:06	78		85		92		99					
14:12	77		84		91		98					
14:18	76		83		90		97					
14:24	75		82		89		96					
14:30	74		81		88		95					
14:36	73		80		87		94					
14:42	72		79		86		93		100			
14:48	71		78		85		92		99			
14:54	70	100	77		84		91		98			
15:00	69	99	76		83		90		97			
15:06	68	98	75		82		89		96	100		
15:12	67	97	74		81		88		95	99		
15:18	66	96	73		80		87		94	98		
15:24	65	95	72		79		86		93	97		
15:30	64	94	71		78		85		92	96		
15:36	63	93	70	100	77		84		91	95	100	
15:42	62	92	69	99	76	100	83		90	94	99	
15:48	61	91	68	98	75	99	82		89	93	98	
15:54	60	90	67	97	74	98	81		88	92	97	
16:00	59	89	66	96	73	97	80		87	91	96	100
16:06	58	88	65	95	72	96	79		86	90	95	99
16:12	57	87	64	94	71	95	78		85	89	94	98
16:18	56	86	63	93	70	94	77		84	88	93	97
16:24	55	85	62	92	69	93	76		83	87	92	96
16:30	54	84	61	91	68	92	75		82	86	91	95
16:36	53	83	60	90	67	91	74		81	85	90	94
16:42	52	82	59	89	66	90	73		80	84	89	93
16:48	51	81	58	88	65	89	72		79	83	88	92
16:54	50	80	57	87	64	88	71		78	82	87	91
17:00	48	79	56	86	63	87	70		77	81	86	90
17:06	46	78	55	85	62	86	69		76	80	85	89
17:12	44	77	54	84	61	85	68		74	79	84	88
17:18	42	76	53	83	60	84	67		74	78	83	87
17:24	40	75	52	82	59	83	66		73	77	82	86
17:30	38	74	51	81	58	82	65		72	76	81	85
17:36	36	73	50	80	57	81	64		71	75	80	84
17:42	34	72	48	79	56	80	63		70	74	79	83
17:48	32	71	46	78	55	79	62		69	73	78	82
17:54	30	70	44	77	54	78	61		68	72	77	81
18:00	28	69	42	76	53	77	60		67	71	76	80
18:06	26	68	40	75	52	76	59		66	70	75	79
18:12	24	67	38	74	51	75	58		65	69	74	78
18:18	22	66	36	73	50	74	57		64	68	73	77
18:24	20	65	34	72	48	73	56		63	67	72	76
18:30	18	64	32	71	46	72	55		62	66	71	75
18:36	16	63	30	70	44	71	54	100	61	65	70	74
18:42	14	62	28	69	42	70	53	99	60	64	69	73
18:48	12	61	26	68	40	69	52	98	58	63	68	72
18:54	10	60	24	67	38	68	51	97	56	62	67	71
19:00	8	59	22	66	36	67	50	96	54	61	66	70
19:06	6	58	20	65	34	66	48	95	52	60	65	69
19:12	4	57	18	64	32	65	46	94	50	58	64	68
19:18	2	56	16	63	30	64	44	93	48	56	63	67

TABLE A-18
APFT Scoring Standards—Two-mile Run
(continued)

Time AGE GROUP	17–21		22–26		27–31		32–36		37–41		42–46		47–51		52 +	
19:24		55	14	62	28	63	42	92	46		54		62		66	
19:30		54	12	61	26	62	40	91	44	▼	52		61		65	
19:36		53	10	60	24	61	38	90	42	100	50		60		64	
19:42		52	8	59	22	60	36	89	40	99	48		58		63	
19:48		51	6	58	20	59	34	88	38	98	46		56		62	
19:54		50	4	57	18	58	32	87	36	97	44	▼	54		61	
20:00		48	2	56	16	57	30	86	34	96	42	100	52		60	
20:06		46		55	14	56	28	85	32	95	40	99	50		58	
20:12		44		54	12	55	26	84	30	94	38	98	48		56	
20:18		42		53	10	54	24	83	28	93	36	97	46		54	
20:24		40		52	8	53	22	82	26	92	34	96	44	▼	52	
20:30		38		51	6	52	20	81	24	91	32	95	42	100	50	
20:36		36		50	4	51	18	80	22	90	30	94	40	99	48	
20:42		34		48	2	50	16	79	20	89	28	93	38	98	46	
20:48		32		46		48	14	78	18	88	26	92	36	97	44	
20:54		30		44		46	12	77	16	87	24	91	34	96	42	▼
21:00		28		42		44	10	76	14	86	22	90	32	95	40	100
21:06		26		40		42	8	75	12	85	20	89	30	94	38	99
21:12		24		38		40	6	74	10	84	18	88	28	93	36	98
21:18		22		36		38	4	73	8	83	16	87	26	92	34	97
21:24		20		34		36	2	72	6	82	14	86	24	91	32	96
21:30		18		32		34		71	4	81	12	85	22	90	30	95
21:36		16		30		32		70	2	80	10	84	20	89	28	94
21:42		14		28		30		69		79	8	83	18	88	26	93
21:48		12		26		28		68		78	6	82	16	87	24	92
21:54		10		24		26		67		77	4	81	14	86	22	91
22:00		8		22		24		66		76	2	80	12	85	20	90
22:06		6		20		22		65		75		79	10	84	18	89
22:12		4		18		20		64		74		78	8	83	16	88
22:18		2		16		18		63		73		77	6	82	14	87
22:24				14		16		62		72		76	4	81	12	86
22:30				12		14		61		71		75	2	80	10	85
22:36				10		12		60		70		74		79	8	84
22:42				8		10		59		69		73		78	6	83
22:48				6		8		58		68		72		77	4	83
22:54				4		6		57		67		71		76	2	81
23:00				2		4		56		66		70		75		80
23:06						2		55		65		69		74		79
23:12								54		64		68		73		78
23:18								53		63		67		72		77
23:24								52		62		66		71		76
23:30								51		61		65		70		75
23:36								50		60		64		69		74
23:42								48		58		63		68		73
23:48								46		56		62		67		72
23:54								44		54		61		66		71
24:00								42		52		60		65		70
24:06								40		50		58		64		69
24:12								38		48		56		63		68
24:18								36		46		54		62		67
24:24								34		44		52		61		66
24:30								32		42		50		60		65
24:36								30		40		48		58		64
24:42								28		38		46		56		63
24:48								26		36		44		54		62
24:54								24		34		42		52		61
25:00								22		32		40		50		60
25:06								20		30		38		48		58
25:12								18		28		36		46		56
25:18								16		26		34		44		54
25:24								14		24		32		42		52
25:30								12		22		30		40		50
25:36								10		20		28		38		48
25:42								8		18		26		36		46
25:48								6		16		24		34		44
25:54								4		14		22		32		42
26:00								2		12		20		30		40
26:06										10		18		28		38
26:12										8		16		26		36
26:18										6		14		24		34
26:24										4		12		22		32
26:30										2		10		20		30
26:36												8		18		28
26:42												6		16		26
26:48												4		14		24
26:54												2		12		22
27:00														10		20
27:06														8		18
27:12														6		16
27:18														4		14
27:24														2		12
27:30																10
27:36																8
27:42																6
27:48																4
27:54																2

TABLE A-19
Spectrum Chart for Electronic Warfare

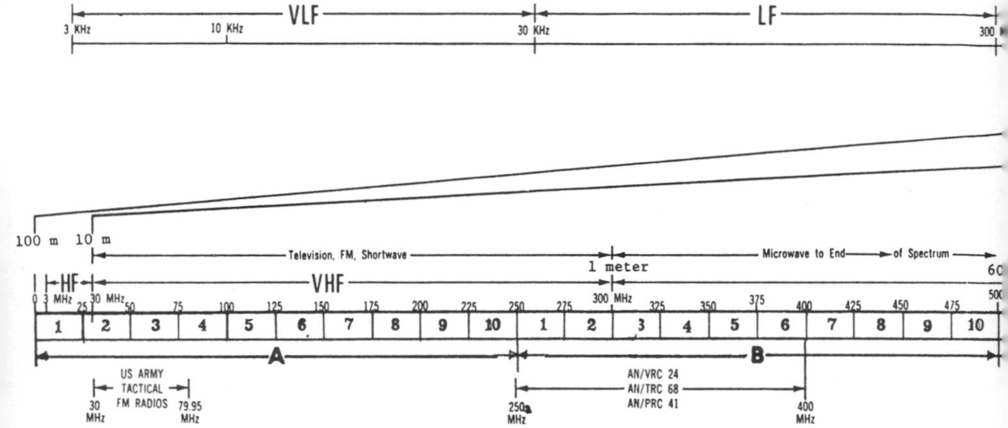

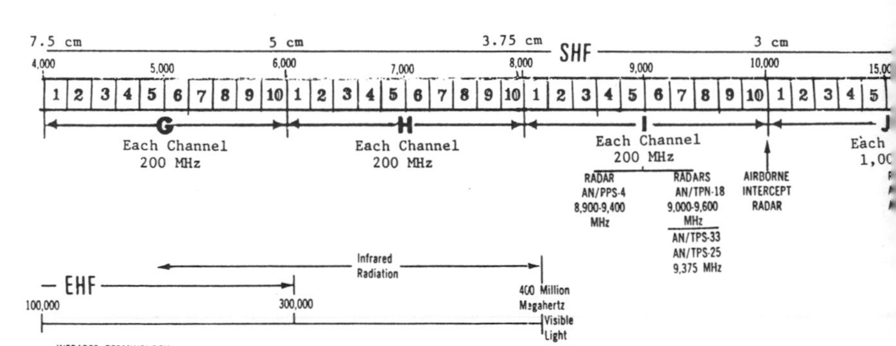

INFRARED TERMINOLOGY

Because, as the frequency of visible light is approached, the numbers representing frequency become so large as to be arithmetically unmanageable, INFRARED designations are expressed in WAVELENGTHS measured in MICRONS. A micron is one millionth of a meter. Since wavelengths decrease as frequency increases, the closer the energy is in frequency to visible light the smaller the micron figure. Technicians and EW writers often refer to IR in terms of its distance from the visible light frequency thus the expressions near, middle, and far IR. The graph below is the approximate coverage of these terms.

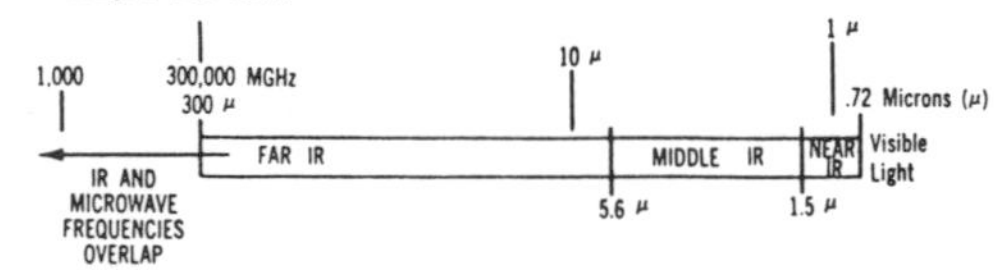

USACGSC 5L5-1332-6Nov74-620

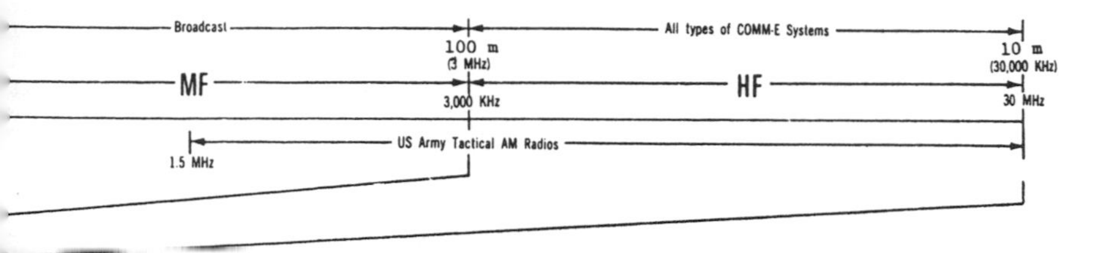

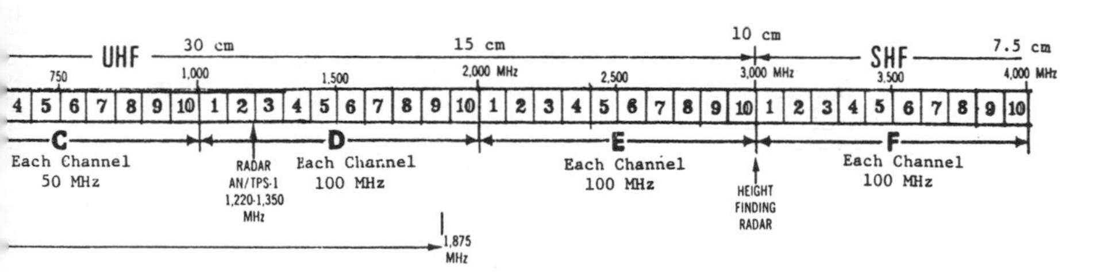

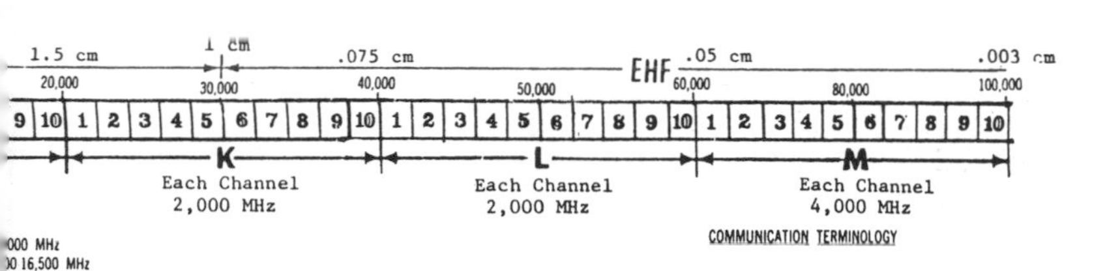

COMMUNICATION TERMINOLOGY

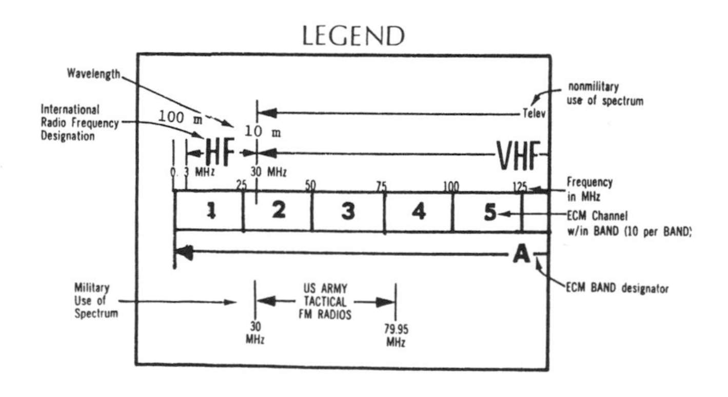

TABLE A-19
Spectrum Chart for Electronic Warfare
(continued)

1 Hz = HERTZ = Frequency of 1 cycle per second
1 KHz = KILOHERTZ = 1,000 hertz
1 MHz = MEGAHERTZ = 1,000 kilohertz
1 GHz = GIGAHERTZ = 1,000 megahertz

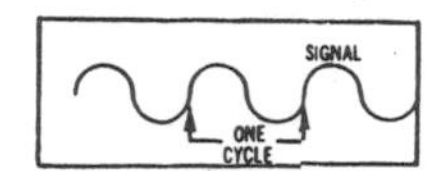

Note: The velocity of electromagnetic waves is the same as that of light = approx 186,284 miles per second.

ECM bands and channels are used to identify spectrum coverage of EW and EW-related equipment.

E Band = 2,000 to 3,000 MHz
E7 channel = 2,600 to 2,700 MHz
E7+50 = 2,650 MHz

FORMER BAND DESIGNATORS

You may come across documents that refer to different RADAR or ECM Bands. The earlier designator are:

G 100 MHz to 225 MHz
P 225 MHz to 390 MHz
L 390 MHz to 1,550 MHz
S 1,550 MHz to 5,200 MHz
X 5,200 MHz to 10,900 MHz
K 10,900 MHz to 36,000 MHz
Q 36,000 MHz to 46,000 MHz
V 46,000 MHz to 56,000 MHz
C 39,000 MHz to 62,000 MHz

ECM bands are shown and lettered. Communicators often use numbered bands when referring to frequency groupings as follows.

COMMUNICATION BAND	INTERNATIONAL RF DESIGNATION	ECM BANDS	FREQUENCY RANGE RANGE
4	Very Low Frequency (VLF)		3-30 KHz
5	Low Frequency (LF)		30-300 KHz
6	Medium Frequency (MF)	Channel ① of A includes	300 KHz 3 MHz
7	High Frequency (HF)	Channel ① plus part of ② of A	3 MHz-30 MHz
8	Very High Frequency (VHF)	A plus Channels ① and ② of B	30-300 MHz
9	Ultra High Frequency (UHF)	From Channel ③ B through E Band	300-3,000 MHz
10	Super High Frequency (SHF)	F through Channel ⑤ of K	3,000 MHz-30,000 MHz
11	Extremely High Frequency (EHF)	Channel ⑥ of K through and beyond M (→100,000)	30,000 MHz-300,000 MHz

Frequency usage is sometimes assigned by wavelength. 6-0770

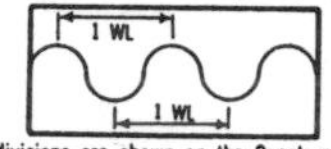

Major subdivisions are shown on the Spectrum in Green.

The following conversion formulas may be useful.

TO CONVERT WAVELENGTH (λ) TO FREQUENCY:

$$\text{Freq (MHz)} = \frac{300}{\lambda\,\text{(meters)}} \text{ or } \frac{984}{\lambda\,\text{(ft)}}$$

TO CONVERT FREQUENCY TO WAVELENGTH:

$$\lambda\,\text{(meters)} = \frac{300}{\text{Freq (MHz)}} \text{ or } \lambda\,\text{(ft)} = \frac{984}{\text{Freq (MHz)}}$$

TABLE A-20
Precedence of Awards, Decorations, and Medals

ORDER OF PRECEDENCE BY CATEGORY OF MEDAL.
The following list indicates the order of precedence by category, when medals from two or more categories are worn at the same time.

a. U.S. military decorations
b. U.S. unit awards
c. U.S. nonmilitary decorations
d. Good Conduct Medal
e. U.S. campaign and service medals
f. U.S. service and training ribbons
g. U.S. merchant marine awards
h. Foreign military decorations
i. Foreign unit awards
j. Non-U.S. service awards

ORDER OF PRECEDENCE WITHIN CATEGORIES OF MEDALS.
The following lists indicate the order of precedence within each category, when two or more medals from each category are worn at the same time.

a. U.S. Military Decorations
(1) Medal of Honor (Army, Navy, Air Force)
(2) Distinguished Service Cross
(3) Navy Cross
(4) Air Force Cross
(5) Defense Distinguished Service Medal
(6) Distinguished Service Medal (Army, Navy, Air Force, Coast Guard)
(7) Silver Star
(8) Defense Superior Service Medal
(9) Legion of Merit
(10) Distinguished Flying Cross
(11) Soldier's Medal
(12) Navy and Marine Corps Medal
(13) Airman's Medal
(14) Coast Guard Medal
(15) Bronze Star Medal
(16) Purple Heart
(17) Defense Meritorious Service Medal
(18) Meritorious Service Medal
(19) Air Medal
(20) Joint Service Commendation Medal
(21) Army Commendation Medal
(22) Navy Commendation Medal
(23) Air Force Commendation Medal
(24) Coast Guard Commendation Medal
(25) Joint Service Achievement Medal
(26) Army Achievement Medal
(27) Navy Achievement Medal
(28) Air Force Achievement Medal
(29) Coast Guard Achievement Medal
(30) Combat Action Ribbon

b. U.S. Unit Awards
(1) Presidential Unit Citation (Army, Air Force)
(2) Presidential Unit Citation (Navy)
(3) Joint Meritorious Unit Award
(4) Valorous Unit Award
(5) Meritorious Unit Commendation (Army)
(6) Navy Unit Commendation
(7) Air Force Outstanding Unit Award
(8) Coast Guard Unit Commendation
(9) Army Superior Unit Award
(10) Meritorious Unit Commendation (Navy)
(11) Navy "E" Ribbon
(12) Air Force Organizational Excellence Award
(13) Coast Guard Meritorious Unit Commendation

c. U.S. Nonmilitary Decorations
(1) Presidential Medal of Freedom
(2) Gold Lifesaving Medal
(3) Medal for Merit
(4) Silver Lifesaving Medal
(5) National Security Medal
(6) Medal of Freedom

TABLE A-20
Precedence of Awards, Decorations, and Medals
(continued)

(7) Distinguished Civilian Service Medal
(8) Outstanding Civilian Service Medal
(9) Selective Service Distinguished, Exceptional, and Meritorious Service Medals
(10) Civilian Service in Vietnam Medal

d. Good Conduct Medal
Good conduct medals from the other services will follow the Army Good Conduct Medal. The Army Reserve Components Achievement Medal will immediately follow, in order of precedence, the Army Good Conduct Medal and/or the Good Conduct Medals from the other U.S. services.

e. U.S. Service (Campaign) Medals and Service and Training Ribbons
Service medals and ribbons awarded by the other U.S. services may also be worn on the Army uniform except the Air Force Longevity Service Award Ribbon and Air Force and Navy marksmanship ribbons. Service and training ribbons awarded by other U.S. services will be worn after U.S. Army service and training ribbons and before foreign awards.
(1) American Defense Service Medal
(2) Women's Army Corps Service Medal
(3) American Campaign Medal
(4) Asiatic—Pacific Campaign Medal
(5) European—African—Middle Eastern Campaign Medal
(6) World War II Victory Medal
(7) Army of Occupation Medal
(8) Medal for Humane Action
(9) National Defense Service Medal
(10) Korean Service Medal
(11) Antarctica Service Medal
(12) Armed Forces Expeditionary Medal
(13) Vietnam Service Medal
(14) Southwest Asia Service Medal
(15) Humanitarian Service Medal
(16) Armed Forces Reserve Medal
(17) NCO Professional Development Ribbon
(18) Army Service Ribbon
(19) Overseas Ribbon
(20) Army Reserve Components Overseas Training Ribbon

f. U.S. Merchant Marine Awards
(1) Merchant Marine Gallant Ship Unit Citation
(2) Merchant Marine Defense Bar
(3) Merchant Marine Combat Bar
(4) Merchant Marine war zone bars
  (a) Atlantic War Zone
  (b) Mediterranean—Middle East War Zone
  (c) Pacific War Zone
(5) Merchant Marine Victory Medal
(6) Merchant Marine Korean Service Bar
(7) Merchant Marine Vietnam Service Bar

g. Foreign Military Decorations
No foreign decorations may be worn on the uniform unless at least one U.S. decoration or service medal is worn at the same time.

h. Foreign Unit Awards
No foreign unit awards may be worn on the Army uniform unless at least one U.S. decoration or service medal is worn at the same time.
(1) Philippine Republic Presidential Unit Citation
(2) Republic of Korea Presidential Unit Citation
(3) Vietnam Presidential Unit Citation
(4) Republic of Vietnam Gallantry Cross Unit Citation
(5) Republic of Vietnam Civil Actions Unit Citation

TABLE A-20
Precedence of Awards, Decorations, and Medals
(continued)

(6) Fourrageres (no order of precedence)
  (a) French Fourragere
  (b) Belgian Fourragere
  (c) Netherlands Orange Lanyard

i. Non-U.S. Service Medals and Ribbons
No non-U.S. service awards may be worn on the Army uniform unless at least one U.S. decoration or service medal is worn at the same time. Other foreign service medals may be worn only if the wearer was awarded such medal while a bona fide member of the armed forces of a friendly foreign nation.

(1) Philippine Defense Ribbon
(2) Philippine Liberation Ribbon
(3) Philippine Independence Ribbon
(4) United Nations Service Medal
(5) Inter-American Defense Board Medal
(6) United Nations Medal
(7) Multinational Force and Observers Medal
(8) Republic of Vietnam Campaign Medal

TABLE A-21
Operational Graphic Symbols

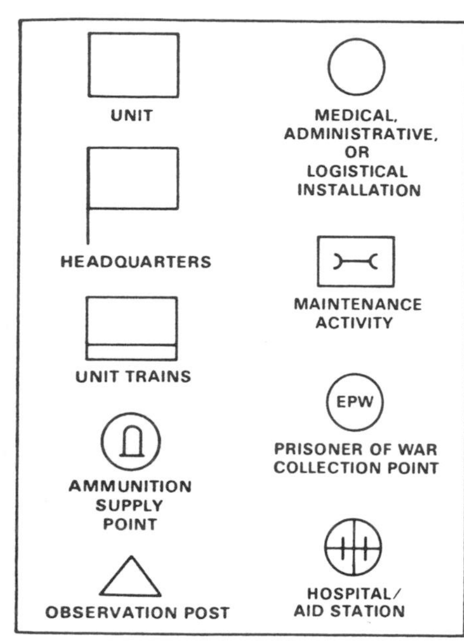

Basic unit symbols

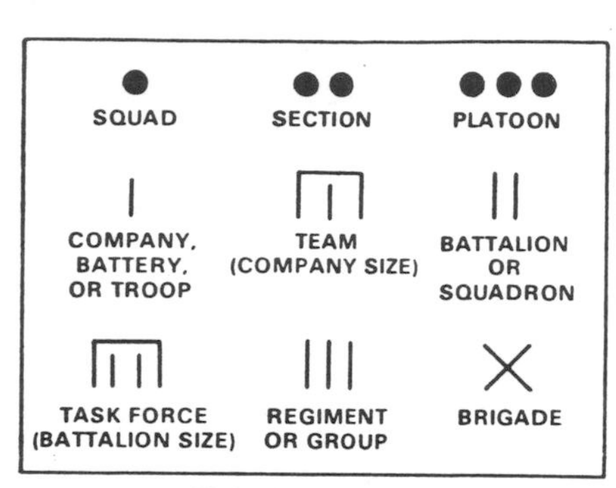

Unit size symbols

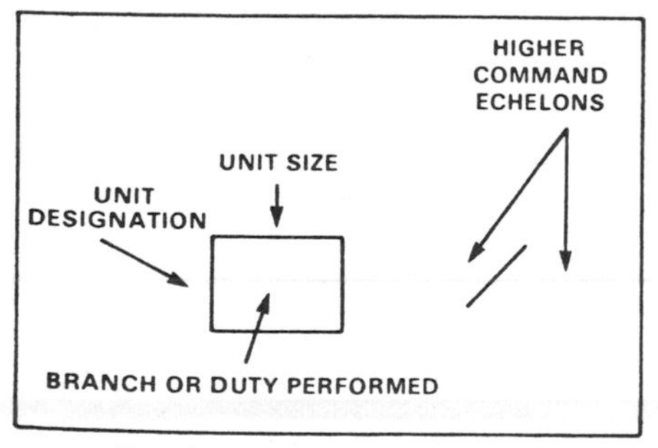

Development of unit symbol

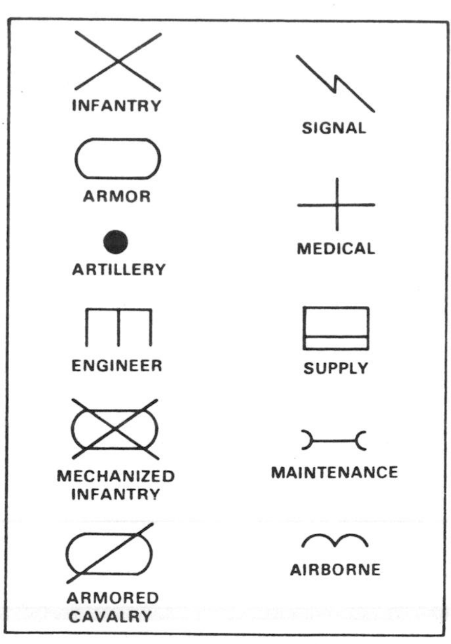

Branch symbols

TABLE A-21
Operational Graphic Symbols
(continued)

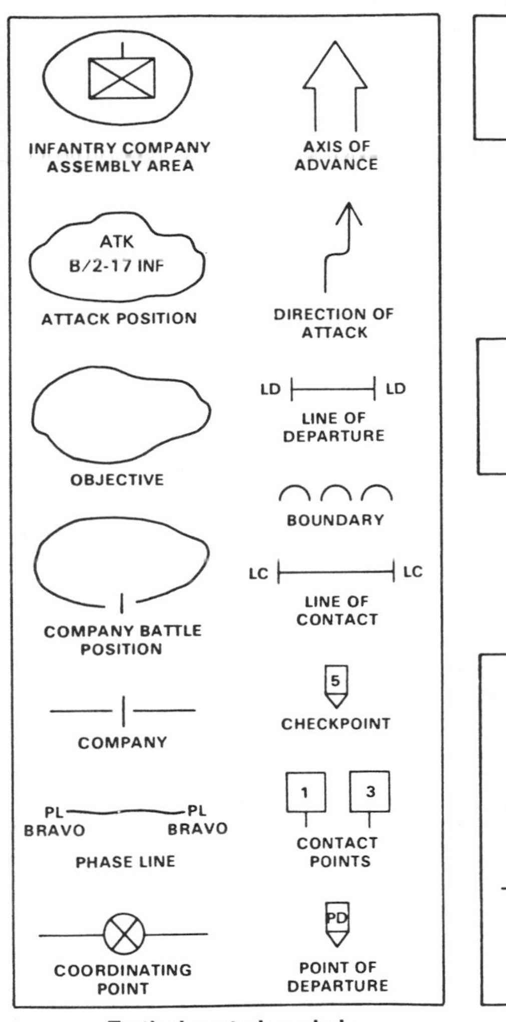

Tactical control symbols

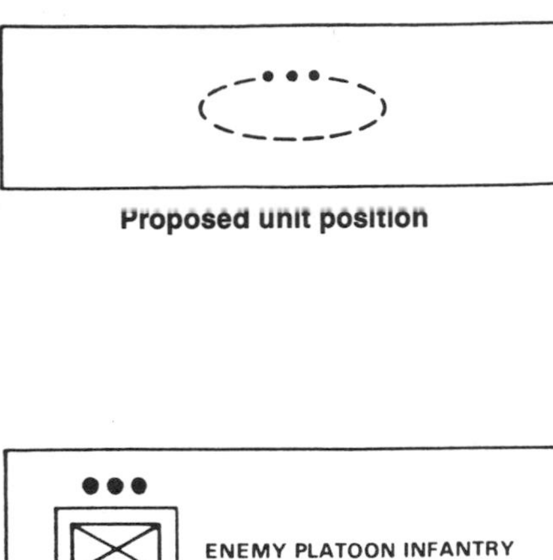

Proposed unit position

Enemy unit

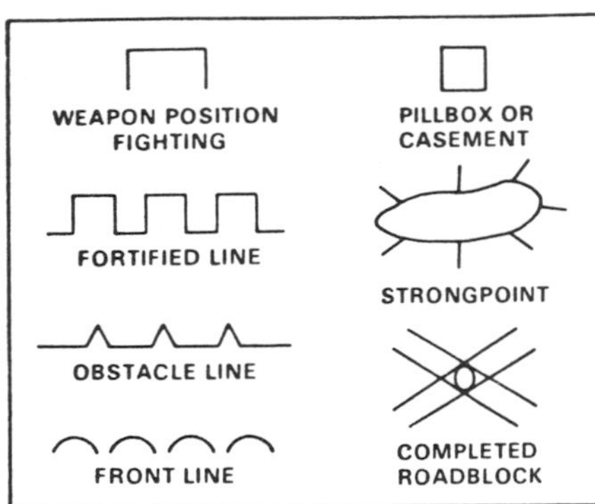

Fortification and obstacle symbols

TABLE A-21
Operational Graphic Symbols
(continued)

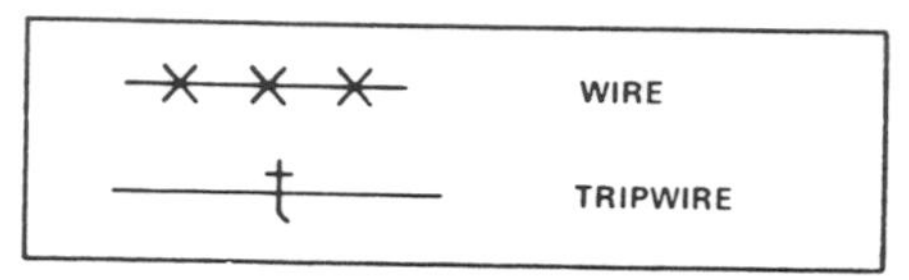

Tactical wire symbols

WEAPON	LIGHT	MEDIUM	HEAVY
AUTOMATIC INFANTRY WEAPONS			
ANTITANK ROCKET LAUNCHERS			
MORTARS	60-MM	81-MM	4.2-INCH
RECOILLESS RIFLES			
ANTITANK MISSILE OR ROCKET			
"EXAMPLE"	(LAW)	(DRAGON)	(TOW)

Weapon symbols

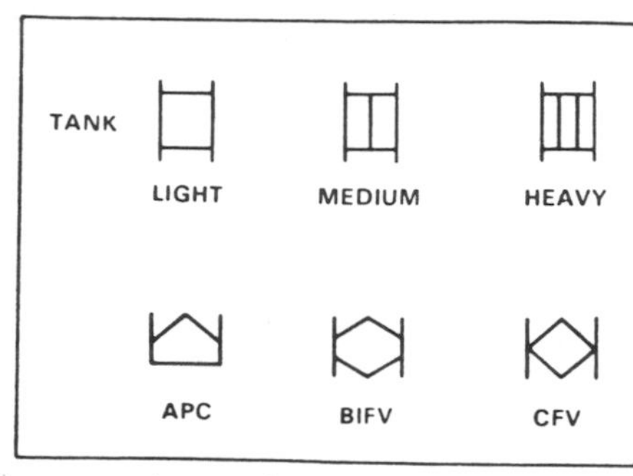

Armored vehicle symbols

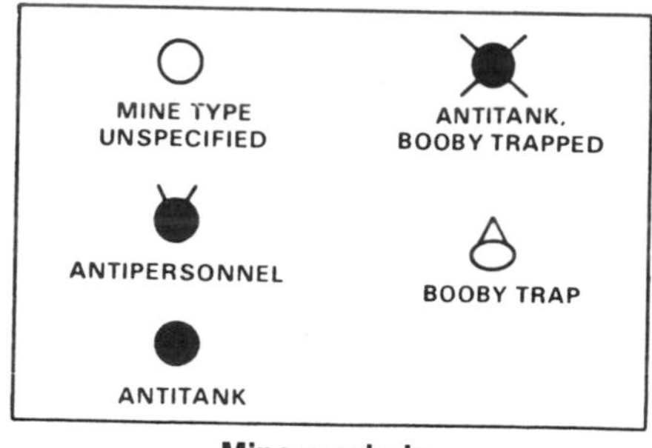

Mine symbols

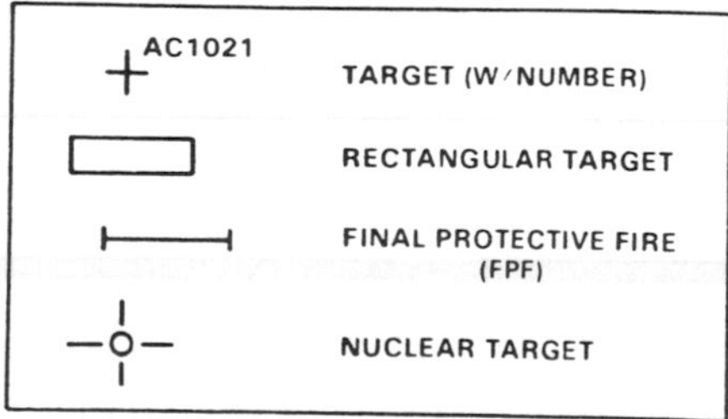

Indirect fire symbols

Code of Conduct for Members of the U.S. Armed Forces

An American soldier's duties include resistance to capture and continue even after his capture by an enemy. These continuing duties are to be conducted within the following framework:

I.

I am an American, fighting in the forces which guard my country and our way of life. I am prepared to give my life in their defense.

II.

I will never surrender of my own free will. If in command, I will never surrender the members of my command while they still have the means to resist.

III.

If I am captured I will continue to resist by all means available. I will make every effort to escape and aid others to escape. I will accept neither parole nor special favors from the enemy.

IV.

If I become a prisoner of war, I will keep faith with my fellow prisoners. I will give no information or take part in any action which might be harmful to my comrades. If I am senior, I will take command. If not, I will obey the lawful orders of those appointed over me, and I will back them up in every way.

V.

When questioned, should I become a prisoner of war, I am required to give my name, rank, service number, and date of birth. I will evade answering further questions to the utmost of my ability. I will make no oral or written statements disloyal to my country or its allies or harmful to their cause.

VI.

I will never forget that I am American, fighting for freedom, responsible for my actions, and dedicated to the principles which made my country free. I will trust in my God and in the United States of America.

Sample Memorandum and Letters

Source: AR 25–50

DEPARTMENT OF THE ARMY
ORGANIZATIONAL NAME/TITLE
CITY, STATE, AND ZIP CODE

REPLY TO
ATTENTION OF

S: SUSPENSE DATE

1
2 OFFICE SYMBOL (MARKS NUMBER) 1 2 DATE
1
2
3 MEMORANDUM FOR Commander, Fifth United States Army,
Fort Sam Houston, TX 78234-6000
1
2 SUBJECT: Preparing a Memorandum
1
2
3 1. This example shows how to prepare a memorandum. Allow one inch for the left and right margins. Use pica, courier, or other 10-pitch type. Use elite, 12-pitch or other available type only when larger type is not available.

a. Type the OFFICE SYMBOL at the left margin two lines below the seal. Type the MARKS Number, in parentheses, two spaces after the office symbol.

b. Stamp or type the DATE on the same line as the office symbol, ending at the right margin. If there is a SUSPENSE DATE, type it two lines above the office symbol line ending as close as possible to the right margin.

c. Type MEMORANDUM FOR on the third line below the office symbol. Begin the single address one space following MEMORANDUM FOR. If the MEMORANDUM FOR address extends more than one line, begin the second line under the third letter of the first word after MEMORANDUM FOR. Addresses may be in upper and lower case type or all upper case type.

d. Type the SUBJECT of the memorandum on the second line below the last line of an address.

e. Begin the first paragraph of the TEXT at the left margin on the third line below the last line of the subject.

2. When used, type the AUTHORITY LINE at the left margin on the second line below the last line of the text.

3. Type the SIGNATURE BLOCK on the fifth line below the authority line or the last line of the text beginning in the center of the page. Identify enclosures, if any, flush with the left margin beginning on the same line as the signature block.

4. Leave at least a 1 to 1-1/2 inch margin at the bottom of the first page.

Preparing a memorandum

6
7
8 OFFICE SYMBOL
1 SUBJECT: Continuing a Memorandum
1
2
3 5. Type the OFFICE SYMBOL at the left margin on the EIGHTH line from the top edge of the paper.

6. Type the SUBJECT of the memorandum at the left margin on the next line below the office symbol.

7. Begin the continuation of the TEXT at the left margin on the third line below the subject phrase. When continuing a memorandum on another page:

a. Do not divide a paragraph of three lines or less between pages. At least two lines of the divided paragraph must appear on each page.

b. Include at least two words on each page of any sentence divided between pages.

c. Avoid hyphenation whenever possible.

d. Do not hyphenate a word between pages.

e. Do not type the AUTHORITY LINE and the SIGNATURE BLOCK on the continuation page without at least two lines of the last paragraph. If, however, a paragraph or subparagraph has only one line, place it alone on the continuation page with the authority line and signature bock.

8. Center the page number approximately 1 to 1-1/2 inches from the bottom of the page.

1
2 AUTHORITY LINE:
1
2
3
4
5 4 Encls
1. Personnel Listing,
24 May 1988
2. DA Form 4187
3. Orders 114-6
4. AR 340-25

SIGNATURE BLOCK
Xxxxxxx, XX
Xxxxxx, Xxxxxxxxxxx
Xxxxxx, Xxxxxxxx

2

Continuing a memorandum

DEPARTMENT OF THE ARMY
ORGANIZATIONAL NAME/TITLE
CITY, STATE, AND ZIP CODE

1
2 July 1, 1987

Manpower Programming
Division

Mr. John A. Doe
123 Main Street
Nashville, Tennessee 73695
1
2 Dear Mr. Doe:
1
2 Adjust margins to where the letter is framed on the page. Use a 10-pitch, common type style.

Type dates in civilian style and center two lines below the last line of the letterhead. Do not use date stamps on original copies. You may use them on file copies.

There is no set number of lines between the REPLY TO ATTENTION OF line, when preprinted on the letterhead and the first line of the address. Frame the letter on the page. Five lines is the general rule when the letter is two or more pages.

Do not use abbreviations in the address or signature blocks.

Single-space the body of a letter with double spacing between paragraphs.

Type the salutation on the second line below the last line of the address.

Type the first line of the text of the letter on the second line below the salutation.

Indent four spaces and begin typing on the fifth for the first line of the paragraph. Do not number paragraphs.

When you need more than one page, there should be a minimum of two lines of text on the second page.

Leave at least a 1-inch margin at the bottom of multiple-page letters.

Basic letter format

3
4
5 -2-

1
2
3
4
5 When more than one page is required, center the page number on the fifth line from the top edge of the paper. Use a hyphen on each side of the page number.

Start the first line of the text on the fifth line below the number of the page, keeping the margins the same as the preceding page(s).

Start the closing on the second line below the last line of the letter. Begin at the center of the page.

Signature blocks will be in upper and lower case. Do not use abbreviations except those authorized in paragraph 4-8.* Military personnel will use "U.S. Army" following their rank. Branch designations and "General Staff" have no meaning to the general public.

Whenever the Secretary of the Army signs on personal letterhead, do not use a title.

You may use either "Enclosure" or "Attachment" as long as it is consistent with the words used in the text. Type the word "Enclosure" or "Attachment" at the left margin on the second line below the signature block. Do not show the number of enclosures or list them. Fully identify enclosures or attachments in the text. When there is more than one enclosure or attachment, use the plural form of the word.

1
2 Sincerely,
1
2
3
4
5 Nathan I. Hale, Jr.
Major General, U.S. Army
Commanding Officer

1
2 Enclosure

Continuation of a letter

AR 25-20

DEPARTMENT OF THE ARMY
ORGANIZATIONAL NAME/TITLE
CITY, STATE, AND ZIP CODE

1
2   July 24, 1987

Director, Futures and Concepts

Honorable Janet R. Wise
Mayor of Woodbridge
Woodbridge, Virginia 22191

1
2 Dear Mayor Wise:
1
2 In a letter to a civilian, office symbols rarely have any meaning and appear awkward. They should only be used on the original letter when absolutely needed. Even here it is better to use an understandable phrase or title, such as "Director of Training," "Comptroller," "Director, Futures and Concepts," if you can do it without appearing awkward. If you need to provide a specific return address, put it in the last paragraph. For easy reference or filing, use office symbols on file copies.

Type office titles on the second line below the seal starting at the left margin. When the letterhead contains four or more lines, type the office title on the second line below the date, flush with the left margin.

Do not use office titles on the original correspondence prepared for the Secretary of the Army's signature.

1
2 Sincerely,
1
2
3
4
5 John L. Ribbons, Jr.
Director, Futures and
Concepts

Use of office symbols and titles

Notes

Notes

Notes

Notes

Notes

Notes

Notes